BUILDING WITH

AND OTHER GOOD STUFF

■

A Guide To
Home Building and Remodeling
Using Recycled Materials

■

Jim Broadstreet

Loompanics Unlimited
Port Townsend, Washington

BUILDING WITH JUNK AND OTHER GOOD STUFF:
A Guide To Home Building and Remodeling Using Recycled Materials
© 1990 by Jim Broadstreet
Printed in USA

Published by:
Loompanics Unlimited
PO Box 1197
Port Townsend, WA 98368

Cover by Technigraphic Systems Inc.

Photo Credits:
Carl Bundschuh: pages 25, 27, 41, 91
Art Evans: pages 26, 31, 36, 40, 49, 57, 86, 114
Arrow Ross: pages 32, 44, 93, 96, 119, 120, 129
Ron Fowler: pages 39, 117
John Rogers: pages 83, 100, 147
The Springfield, MO, *News-Leader*: page 116
All photos courtesy of Jim Broadstreet

ISBN 1-55950-036-0
Library of Congress
  Catalog Card Number 90-061243

Contents

Acknowledgements 1

1. Introduction 3

2. Obtaining and Storing Materials 5

3. Building Codes and Regulations 13

4. Financing 17

5. Concrete (and Alternatives to Its Use) 23

6. Masonry (Blocks, Bricks, Stone, Rocks, Marble, Adobe) 29

7. Wood-Framing Lumber and Large Pieces 35

8. Wood-Finish Lumber and Small Pieces 43

9. Millwork and Cabinets 51

10. Plywood, Particle Board, Fiberboard and Hardboard 63

11. Roofing Materials 71

12. Insulation, Vapor Barriers and Sealants 77

13. Metals 81

14. Glass and Mirrors 85

15. Plastics 89

16. Doors and Pieces of Doors 95

17. Hardware 101

18. Windows 105

19. Wall and Ceiling Materials and Finishes 109

20. Floor Coverings 113

21. Furniture and Art 117

22. Appliances 123
23. Electricity and Lighting....................................127
24. Plumbing 131
25. Heating, Air-Conditioning and Ventilating 139
26. Uses of Solar Energy 145
27. Tools and Equipment 149
    Index...157

■

Acknowledgements

Over the years, I have been fortunate to meet many wonderful, interesting people who have the ability to think abstractly and have an interest in scrounging and making-do with what's available. Some, like Keith Freeman, enjoy accumulating more than using — thus his place is always a good source of supply and point of inspiration. Others, like Henry Pina, in Portales, New Mexico, can use stuff faster than it's possible to acquire — so his work is also an inspiration.

During the writing of this book, I was associated with Bob DeLong on several projects where I did the design and he the building. Usually it is Bob I have referred to when using the word "we." Bob is a damned good scavenger *and* builder.

Some dear old friends encouraged me to write and allowed me to have pictures of their accomplishments with junk and other good stuff. Architect Ed Waters incorporated the technique in the building of his lake house, which is mentioned several times. Artist Beverly Hopkins, also mentioned, did fabulous things with a weird assortment of finds. Dave Karchmer used to call me when he found gems for which he had no immediate use in one of his own wonderful creations. Architects John McWilliams and Dennis Spencer bought a bunch of solid-core doors from me to create their new office furnishings and then to expand it over the years as they became more and more successful. My son Jim has created

the furnishings for his old, catalog-ordered house with delightfully bizarre good taste.

Max and Charlene Skidmore let me have some fun with their kitchen walls, and she was always working magic with her own finds. (Jim and Dalene Vanderford allowed me to get photos of their wonderful old house after they had acquired it from the Skidmores.) Rick and Theresa Gilmore let me hang their house out over a valley and into the treetops. Harlan and Glenda Jones purchased one of "my houses" from Richard and Sarah Turner and contributed the photograph of the slate-blackboard floor. Several of my buddies are also masters of photography and captured exactly what I asked them to. Photographer Art Evans is also a scrounger/builder of some note. Others include my son Jim, Arrow Ross, Carl Bundschuh of Munich, Germany, and John Rogers, who is also a creative cook.

Judy Davis, bless her, deciphered my handwriting and typed most of the manuscript. My sister Anne Treadway always egged me on and typed afterthoughts.

And throughout everything, my living-mate Velma endured stuff being hauled in and out, peculiar-sounding phone calls, and my child-like fits of excitement over discovered "treasures," etc., etc.

I know I am a blessed man to have so many talented, fascinating, non-conformist friends, only a very few of whom I have mentioned here.

■

CHAPTER ONE

Introduction

An elaborate and rather rigid system for building homes has evolved in our society. All of us who would build are almost obliged to adopt this system as our own. The more obedient to this system we become, the less variety is available, the less creativity is expressed, the less we are able to change with changing times.

The author prefers the taste of poke greens scrounged from among the weeds to store-bought spinach, and houses which express the owner's individuality to rows of ticky-tack.

While it is true that not enough poke greens grow wild to supply everyone, at the present time most of them go to waste. Not everyone has the ability to create his or her own shelter, but there are millions who could, using materials others leave behind. Instead, most people go with the system, waiting years to pay more money to move into something mundane.

This book is for those who will break from the system, who will use their brains and not their billfolds to create homes that are affordable, individual and even safer and more efficient than the mass-produced mundanities of the established construction industry.

There would not be enough "junk" to go around if everyone got into the act. But there is certainly more than enough for those of us who will use it. The United States and Canada throw away more materials every year than most countries produce.

Billboards advertise "Custom Designed, Energy Efficient Homes." Almost always, they are houses mass-produced from the same floor plan, with no more than average insulation, which must be painted different colors so the individual owner can give directions to people trying to find their house.

This book is not intended to explain *exactly* how to build outside the system. Rather, it aims at encouraging expressions of individuality in design and innovation in building methods. If it offers ideas you can use or build upon — fine. If it stimulates *you* to come up with new and better ideas — wonderful!

There is absolutely nothing ungodly, unpatriotic or immoral about living in a creation of one's own. Indeed, you will do great good when you recycle materials into a shelter which someone will find pleasant and in which you may live satisfied and comfortable, using only minimal amounts of the earth's resources.

■

CHAPTER TWO

Obtaining and Storing Materials

This book will attempt to give inspiration and guidance to anyone wanting to save money building.

Building materials, regardless of where they are obtained, fall into six categories:

1. Conventional new materials used conventionally.
2. Conventional used materials used conventionally.
3. Conventional new materials used unconventionally.
4. Conventional used materials used unconventionally.
5. Unconventional new materials.
6. Unconventional old materials.

Materials can always be purchased, in the conventional manner, at lumber yards, hardware stores, plumbing supply houses, etc. When you build, you will doubtless obtain some things you need in the conventional manner.

Conventional building materials are almost always available from sources other than the normal, retail outlets. Prices are often substantially lower, and sometimes drastically so. But don't just assume you will pay less if you get off the beaten path.

To take advantage of good buys, you will need to put yourself in an advantageous position. Most importantly, you will need to learn a good bit about going-rate prices of building materials from normal sources in your area.

This is important, obviously, as protection against being ripped off by someone who "saw you coming" or by you, yourself, caught up with buying fever and agreeing too quickly to an asked price or bidding too high.

■

It will be necessary for you to have someplace secure to store your finds until you can use them. Some types of building materials can be kept outside, exposed to the elements. Some need only to be protected from rain and snow, so can be covered or wrapped. Polyethylene plastic sheeting is relatively inexpensive and comes in some enormous sizes, such as 100-feet rolls, 20′ wide (the width is folded several times before it is rolled). Four-mil and 6-mil thicknesses are readily available. Thicker sheets are also manufactured, but they are generally utilized for things other than wrapping.

This thin stuff is used for other than wrapping purposes also. So don't throw it away after its initial use. You will need it for vapor barriers on the ground in the crawl space below the floor (see chapter on insulation); temporary protection against wind and blowing rain at unfinished walls, unglazed windows, etc.; hanging sheets between finished and unfinished areas to protect finished (or cleaned) areas from construction dirt; drop cloths to protect floors, cabinets, equipment, etc., from dropped paint and sheetrock mud, or plain-old mud being tracked in, etc.

Poly plastic sheeting is available in clear and black. While clear is obviously better for temporary window covering, we recommend black for wrapping lumber, insulation and most building materials, because it blocks direct sun rays.

Four-mil plastic is *extremely* thin. It should sometimes be used double or triple-thickness. It is also somewhat subject to ripping when it is punctured by a corner of a board, the end of a pipe or something. For this reason, it is important to tie, strap or nail the covering on to protect against the wind making it flap, puncture and tear. In nailing, it is advisable to roll the edge to be secured around a board, pull the whole thing tight and nail through board and all. Nailing only through the plastic probably won't even last long enough for you to walk away.

Other materials you will accumulate should be better protected. Doors, windows, cabinets, carpet, interior-grade plywood, paneling, hardwood lumber, particle board, nails, sheetrock and ceiling tiles need more protection from moisture than you will ordinarily get from poly plastic. Find an unused garage, barn, house, store building, etc., and, unless it is a weathertight structure with unbroken windows, throw a piece of poly over the stock to protect it from birds and dirt.

Some goodies you find will need to be protected from freezing. Things like glue, paint and caulking, for instance, will have labels warning against freezing. Remember, if you find some of this type of stuff cheap, it *might* be cheap because it froze while the other guy had it!

Some other items you will want to secure against thieves and vandals. It's a disheartening experience to find a gem like a stained-glass window or a set of crystal door knobs, plan your house around it and then lose it. Is there any room left under your bed?

One more thing about storing materials in stacks: Because of the differences in moisture content and temperature inside and outside of a stack of doors or lumber, let's say, the top door or piece of lumber will warp (curl up) alarmingly. By covering the stack with a piece of weighted-down plywood, this can be avoided. If you do have this problem, simply turn the top piece over and leave it that way until it is straight. A door which is hung when it is warped will probably stay warped forevermore.

■

Another thing you will need, besides knowledge of prices and storage facilities, is some money. How much at any one time depends on how fast you want to accumulate your stock.

If you have a storage place and patience, you might allocate a portion of every paycheck for "house stuff." A disadvantage to this is that you might miss a hell of a deal on a bunch of stuff because it sells for more than you have that week.

The best situation is simply to be rich. But rich folks buy things professionally made and miss all the fun. Next best is to have a reasonable amount laid by or available to allow you to get out and search in earnest. Sometimes this means borrowing with an open line of credit. This type of thing is discussed in the chapter on financing.

We have seen a lot of projects begin before much material was acquired, and some of them have turned out fine. Some have been interesting and some disasters. To build this way one must either wait until he can find what he needs next or modify his plan to suit what he finds. The latter produces an "interesting" structure.

An old Ozarker once told us that you need to "build your outhouse accordin' to your boards." How does that jibe with Louis Sullivan's statement that "form follows function?"

■

Another advantage you can give yourself is to know what can and cannot be used to meet code requirements in your area. The chapter on codes addresses this subject.

You don't want to spend a bunch of time and money collecting materials only to find you are not allowed to use them.

∎

A pickup, van or flatbed truck will allow you to take advantage of deals and "throw-aways" when they present themselves. A station wagon is next best, but you'll sure tear it up before your project is finished.

∎

Besides retailers' regular "overstocked" or "fire" sales, there are often "going-out-of-business" and "stock liquidation" sales. Our advice is to nose around a little and find out how legitimate the sale is. Often, if the sale is scheduled to go on for a considerable length of time, the firm will reduce prices somewhat and even replenish items which sell well. Again, you must know going-rate prices.

There are "going-out-of-business" sales when a firm is closing out one of its branch outlets, in which case it will take anything it doesn't sell at the sale to another branch. The "dog" items will be rounded up from other branches and brought to the one that is closing and conducting a sale. This is fine! One person's "dog" might be another's gem, and the doggier it is (the longer it has set around the place) the cheaper the asked price is likely to be. Know what's a bargain, however.

Bankrupted businesses are sometimes taken over by holding companies, banks or "receivers" to run and/or liquidate, with the hope of salvaging more than would be possible just putting the whole thing on the block at auction. Know prices and watch these. Banks don't know much about building materials and make some unbelievable mistakes.

∎

Auction sales can be one of the very best sources of supply for your project. If you don't know going-rate prices of things, however, they can also cost you bucks you shouldn't have spent that way. Another word of caution: they can be as addictive to some people as alcohol is to others!

If you have never been to an auction, you might tend to be afraid of them because you've heard stories of people being tricked or "sold" something because someone scratched his nose. Believe me, this is not a problem at the auctions we are going to discuss here. (You would not be likely to find what you need at a Sotheby sale.) Occasionally something a little unethical will happen, but there's nothing anyone can do about it. One such practice is that of a "bottom dollar" being placed on an item by the owner. (If he's going to do that, he should say so in his ad in the paper.) Another unethical practice is for the auction company to have someone planted in the crowd to bid an item to a predetermined amount (which he knows he can get for it at another sale) before he quits bidding and lets the crowd take over. These bidders are usually spotted by someone who is a steady auction attender and the word will get through the crowd. Keep your ears open for this, and if you suspect something, ask some old-timer standing near you.

One thing that all auctions have in common is the number given each person when he signs up. The number is shown to the auctioneer's assistant by the "successful" bidder, to be recorded in the "sold" book.

The types of auctions we will discuss here vary enormously, from tiny ones that take about an hour start-to-finish, to huge ones that may span three days, and from ones which have an item or two you could use for your building project, to those that have building materials almost exclusively.

As mentioned before, businesses of all kinds, including building materials retailers and wholesalers, can be foreclosed upon, can be declared bankrupt, or can just run out of steam and quit. An auction is a fast, clean (cash-on-the-barrelhead), sure way of disposing of the saleable merchandise, as well as construction equipment, office equipment, supplies, trucks and sometimes even the real estate. The total receipts will, however, be a gamble.

An auction of this kind very naturally draws people interested in buying building materials and, when a lot of people want the same item, the bidding tends to go high. Other dealers, for instance, can pay close to wholesale price, save freight costs, and make a profit. So real, honest-to-Pete bargains may not be there on the more common items. But remember: most people there are not very creative or innovative, and many are buying for re-sale. So off-beat, out-of-production, damaged, incomplete or mixed-lot stuff might go very reasonably.

When very large businesses are sold at auction, buyers from other businesses of the same type attend, some coming long distances. Except for the off-beat stuff, the merchandise will be bid in such large quantities that the one-project builder can't compete.

Occasionally there is a *big* auction of building materials, such as when a general contractor goes broke on a big project. If you are not in need of a 250-ton bridge crane or 12-yard concrete batch plant, you can forget these. If you are interested in items of that nature, you've picked up the wrong book. If you happen to live near one of these auction sites, you could probably pick up enough pieces of framing lumber and miscellaneous items off the grounds, discarded, to build a house.

But when a home-town contractor or house-building firm is auctioned, there may be all kinds of really neat stuff. Often a contractor will transport removed material from a renovation or expansion job to his yard because "it's just too damned good to throw away." For the life of the company, the yard is where the stuff will inevitably remain. There should be gobs of used stuff, from form-plywood to door closers, and batches of all kinds of new stuff, either in quantities too small to be returned or too dirty or scratched to return for credit. Don't miss these auctions! Brush up on what new stuff is

selling for and go get a bunch for a small fraction of the new price.

Auctions at non-building related businesses sometimes produce usable material, but you might have to wait hours for it to be sold. Sometimes the auctioneer or an assistant will tell you about when something will be up for sale, so you can put your time to good use. Store fixtures, shelving, adjustable shelf hardware (standards and brackets), clothes rods and mirrors usually sell for near nothing. We once picked up a bunch of rolling display units for two dollars apiece. Each yielded four four-inch rubber-wheel casters (just price *them* new sometime), shelf hardware, glass shelves, four-foot by four-foot pieces of ¾″ plywood and ¼″ pegboard and some lumber pieces.

Farm auctions usually include some good stuff. Like stacks of corrugated metal roofing, boards, pipe, etc., and occasionally a pile of hardwood lumber such as walnut, oak, ash, cherry, or gum which the old farmer had cut and stacked back in '39 and always intended to make into something "real purty." He would want you to do it for him now. You might want to pick up some old tools, but be careful — they usually go high. Look around for unusual items that you could find a use for. Many old farmers just couldn't stand to throw anything away. And watch for cast-iron stoves, foot scrapers, fans, etc.

City and town auctions (either estate auctions or those brought about by the owner's moving away) are of various sizes and degrees of interest. Often you can go early, or the previous day, to see what goodies there might be. Look for the same things mentioned above. A friend once bought "everything back in the crawl space off the partial basement." There were styrofoam planks, hardwood boards, old 2 x 4's and 2 x 10's, boxes of old hardware, rolls of screenwire and poly plastic sheeting, and an old tackle box which contained (among other things) a few old solid-silver coins. It was the last lot of the day and he got it for his opening bid of three bucks!

Remember: research prices and keep lists if you don't have a real good memory for such things. You should also get a recent Sears-Roebuck-type catalog and take it with you. You will be amazed how many people will ask to look something up in it!

∎

Real estate is sometimes sold at estate auctions with everything else, and sometimes (in cases of unpaid property tax) "on the court house steps."

If you do not yet have a building site or if you are interested in finding an old place and fixing it up (to live in or for rental property) with bargain materials, you just might catch a "sleeper" at an auction.

Commonly, these properties are pretty run-down and in less-than-fashionable neighborhoods. It is always best to find the worst house on the block and fix it up. It is, from a resale-value point of view, foolhardy to buy a place and make it into the best one in the area.

It might even be possible to acquire a property so inexpensively that you can tear it down for its materials and sell the empty lot for a profit.

Buying real estate at an auction requires that you have cash or a line of credit. We address this in the chapter on financing.

∎

Yard sales don't often offer much, but it doesn't hurt to keep an eye out.

∎

Watch the newspaper for special types of sales such as "as is," "parking lot," "white elephant," etc., at businesses which deal in building materials and related items. They can't afford to warehouse stuff that doesn't sell.

∎

Finding materials which are free is even more rewarding than finding bargains. Now wasn't that a brilliant statement? Really, to find some good stuff someplace that costs you only hauling it away can lift your spirits as much as a love letter.

We do not want to become nuisances or beggars, but it wouldn't hurt to become a "scrounger" if you aren't one already.

Building demolition sites are good places to scrounge, some much better than others.

When structures are being totally removed to make way for "progress," the people doing the removing are probably in one of these two categories: demolition wrecking companies or salvage wrecking companies. Note that they are both "wrecking companies," so you can't tell by the sign on the side of the truck which kind one is.

To find out, you can inquire (ask the guy driving the truck, or better yet, call the company office) or you can wait and see what happens. But when the wrecking ball starts to swing or the bulldozer has made contact with the structure, it is a little too late.

Some wrecking companies handle both salvage and demolition. They will send a crew inside, around and over the structure to remove those items they know will produce a nice, fast profit, and then they will lower the boom.

Often, if you have prior knowledge that something is to come down, you can contact the owner and obtain permission to remove some things. By this time the owner is not usually the last occupant or landlord, but someone like the highway department, Land Clearance for Redevelopment Agency, park board or church.

Remember that the early bird gets the worm, so listen carefully for announcements and rumors. It won't hurt to call.

Companies that send a crew in to dismantle a house or building pretty well know the value of stuff, so you won't find many "steals" from them, but they are certainly good sources of supply. If one of these companies sells everything from the demolition site (which many do), it will have a deadline of getting the stuff away from there, and it will not have to retrieve hauling expense.

A company with a "yard" and storage building will usually take its goodies there for inventory and sale. Generally these prices are higher but, even there, the price of something tends to drop lower and lower the longer it sits unsold.

Wrecking companies do not only demolish houses and small buildings. You have probably seen films of multi-story buildings being dynamited into a heap without any damage to adjacent property or even bricks flying out onto the street. Abandoned car-traffic bridges and railway trestles are put out for bids to wrecking companies. Sometimes the bids are "minus-figure" (the owner pays the wrecking company) and sometimes "plus-figure" (the wrecking company is willing to pay something because it is going to salvage enough to turn a profit). Bridges and railroad trestles yield steel beams, plates, channels, angles, bolts, cables, anchoring devices, grates, pipes, wood members from 2 x 4's to 16 x 24's and larger, cut stone, railroad ties, etc.

How are wrecking companies and Mexican public markets alike? Whatever your answer, it's probably correct, but here is ours: each require that you dicker to get the lowest possible price!

Another thing — items are often not price-marked (and when they are, the price is what the guy *wishes* he could get for it), so wander around and look at everything with a poker face and don't exclaim about how interesting or fine anything is. Ask the price of many items, including (dispersed out in the middle of the list) those you have a genuine interest in. Then make counter-offers of less than you would be willing to pay, etc. If you appear to be someone who might need several items now and come back later with friends, you might get better deals.

These places don't tend to be very sophisticated, and they are not likely to have comput-

ers in use. How much the proprietor has in an item, based on his cost of the structure, labor to tear it down, remove nails and haul it, is virtually impossible to compute, so the price may be more in accordance with how much he likes it. His likes and yours may vary widely.

■

Old factory buildings yield incredible things! Huge boiler tanks, steel doors, steel posts, beams and bar joists, water towers and tanks, wire mesh enclosure-walls, bins, benches, wire-glass, wood of all kinds, bricks and on and on.

Cotton gins are now coming down throughout the South. Warehouse buildings, smoke stacks, railway roundhouses and mining company buildings are being removed.

In the decaying center-city areas there are storage buildings, hotels, theaters, office buildings and restaurant/bar buildings slated for demolition — many because they have become havens for destitute people (thus extreme fire hazards) or the roofs are beginning to leak so badly that they will soon rot away and fall in. Generally, too, taxes are much less on a vacant lot.

We have certainly not mentioned all of the types of structures scheduled for demolition. When you see or hear of one, sit down and think of all the things that might be in it. It is a rare structure, indeed, that won't yield *any* goodies!

■

The United States Government must surely be the world's worst trader. Somehow it has latched onto the idea that the way to trade is to "buy high and sell low."

Consequently, sales of government-owned merchandise offer some astounding bargains on everything from used equipment to "reject" merchandise to top-quality new stuff, in its original packaging, of every conceivable type of product. Because no governmental entity wants to have its budgeted allocation cut, each one will spend every penny it has before the fiscal-year deadline, even if the requisitional merchandise will be surplus at the time of delivery or will replace items which still have 95 percent of their usefulness ahead of them. Then follow various types of "surplus commodity sales."

This phenomenon occurs most frequently, it seems, at military installations, where any imaginable building materials are likely to be found.

A friend recently purchased a commissary building at an air base for $25. This building started out to be a plain metal building but an "over-building" had been built to hide the ugly "tin" structure.

This metal building was disassembled and sold to another party, re-erected, for $10,000, which covered a major portion of the cost of the entire salvage project.

The lumber salvaged from the outer building included 12,000 2 x 10's thirty feet long and number one douglas fir! Other goodies included four 330,000 BTU heaters (two of which had never been lit), forty-eight strands of #3 copper wire 200 feet long each, a 12' x 14' walk-in cooler complete with compressor, another 7½ ton compressor, 160 8' fluorescent light fixtures, 21,000 feet of suspended ceiling, 400 sheets of ¾" marine plywood, etc., etc., etc.

On the minus-side — workers wearing masks removed burial-type insulation (asbestos), bailed it in plastic sacks and transported it (after approval from State Transportation Department) to an EPA approved disposal site. This was a two-day operation which cost him a total of $350 plus considerable time to cut through red tape.

Another type of demolition project involves buildings which are being remodeled, refurbished, renovated and/or added onto. These can be drastically different from those mentioned above. Since demolition is only an incidental portion of the project, there is usually no wrecking company involved. For one thing, union contractors cannot easily hire non-union wreckers. This means that the general contractor will send some of his laborers and a few carpenters, plumbers, tinners and electricians in to "get rid of that stuff so we can get on with the job."

To a contractor, time is money and there is no way it could be profitable to take extra time to salvage something and, heaven forbid, try to sell it. The idea is to pull it out, throw it on the truck and drive it to the dump.

We recently heard about a portion of a building which had been a warehouse being converted into more showroom area. Inquiry revealed that the suspended acoustical-tile ceiling was coming out, as well as a bunch of ductwork, grilles, etc. We are now in the possession of several hundred 2' x 4' pieces of 1" vinyl-covered fiberglass tiles, a few 2' x 4' "egg-crate" light lenses and the "T-bar" grid system in which they were hanging (as well as other miscellaneous items). Because the tiles are lightweight, many of them were not damaged at all when dropped on the floor. And those that cannot be used for a ceiling still have an insulation value of about R-4 (see chapter on insulation), which is pretty damned good stuff. The grid system came down almost intact because it was taken down by craftsmen who usually put it up. They unsnapped each piece and let it fall instead of yanking it down and bending it all up.

We were told that four dumpster-loads had been hauled away before we got there and it had cost the contractor $85 per load for dumpster rental and hauling.

Because removal-from-site costs money, it makes sense that if a contractor can get someone to take it away for him, he will end up with more profit.

However, other factors sometimes come into play. Some contractors are concerned about their (or the owner's) liability and what the insurance company would say. Others are concerned that a salvager will get in the way of the workers or will stand around and talk with $15-an-hour employees. There are, also, some miserable bastards who would rather destroy something than let anyone get any good out of it.

But it's worth a try. Be polite, business-like and pleasant. If enough stuff is involved, it might even be worthwhile to strike a deal whereby you will haul all the junk away for the good stuff in it; but beware: there will be about five times more trash than you imagine. Or, you might consider signing some sort of "hold harmless" agreement with the contractor and/or owner whereby you, alone, would be responsible for any injury you might sustain on the job site.

■

Many items are shipped from plants all around the world in wood crates and on wood pallets. And rail cars and trucks are often loaded with wood restrainer braces to prevent loads from shifting or falling. The lumber used for these purposes ranges all the way from bad stuff to finish lumber.

We have seen crates made of mahogany (originating in the Orient), maple (from Canada and New England), beautifully grained tupelo, willow and brown ash (from the southern states) and exotic woods from South America and Africa.

Bracing pieces are often good-grade 2 x 4 lumber.

The big disadvantages with this material are these: it is often "green," meaning it has been neither air-dried or kiln-dried; it can be full of nails and screws (the screws can often be salvaged); and it is sometimes rough-sawn and irregular in thickness. More about this in the chapter on "Wood-Finish Lumber and Small Pieces."

Lumber is not all that used packaging can yield! Foam pads, styrofoam molds and pellets, steel straps, corner braces, plywood, hardboard, nuts, bolts, washers, etc., can be found in abundance and put to good use. See chapters on "Insulation," "Plastics," "Metals," etc.

There is little need to try to figure out who would have shipments arriving in packaging containing wood, because almost everything imaginable comes that way. A trip through the industrial districts, warehousing areas, retail and wholesale districts (especially along railways) will indicate what's available.

Railroads have "team tracks" situated around in various locations for people to use when they receive a rail shipment but have no rail facilities of their own. Calls to your local railroad companies will get you the location of these. They often provide blocking and bracing lumber.

Glass is shipped in crates made of very thin plywood stapled to edge frames. The chapter on plywood explains a little about this stuff and possible uses.

Glass companies throw crates away all the time and will be glad if you haul them away. But ask first!

■

This chapter has certainly not covered all sources of materials or ways to store them! When you start accumulating and develop a keen eye, you will discover stuff in incredible places and under amazing circumstances! You will probably begin to see possible available storage areas also.

The various different chapters will give ideas on where and how to locate various types of materials.

■

CHAPTER THREE

Building Codes and Regulations

The subject of building codes and regulations is not an easy one to approach in a book aimed at creative, free-spirited people interested in "beating the system" by creating something good without following prescribed norms of procedure.

Many harsh words have been written about building codes, and the people who interpret and enforce them, by people who have built unique structures and have gone on to write about their experiences.

While we concur with most of these derogatory blasts, and are in sympathy with almost all of them, we know there is another side of the coin. We also feel strongly that there is usually a better way than head-on, bitter confrontation with potential adversaries.

It is absolutely true that building codes have often saved lives, protected people from injuries, made life easier for the handicapped, and sometimes, helped preserve our environment. And codes have done some good in protecting unknowledgeable new-home buyers from being screwed by unscrupulous builders. We acknowledge these positive code accomplishments.

Remember, too, that building codes are primarily concerned with life and safety protection in public and quasi-public buildings and most are not applicable to single-family residential construction. Yet.

Some code requirements which do pertain to houses are of extreme importance, and it would greatly benefit all to know about them and understand them before they light into building projects.

If you intend to perform all labor on your project yourself, including electrical, plumbing, heating/air-conditioning and ventilating, you may do so legally. No code can take away your right to be your own carpenter, plumber, painter, electrician or anything else, when you confine this work to your own private residence to be occupied solely by you and your family.

There are, however, some regulations which you must, legally, comply with. Most cities, towns and counties require that you obtain a building permit and call for final inspection of your electrical, plumbing and heating systems to insure that the work you have done will not, for instance, result in a fire that might destroy the neighborhood. Usually, you cannot get hooked up to utilities (electricity, gas, water and sewer) until an inspection has been made and your work has been determined to be safe and in compliance with local codes.

The same inspection requirement is imposed on licensed tradespeople. There have been many complaints that inspectors are much "tougher" on the home owner/builder, and this is probably often the case. Perhaps the inspector *should* examine your work more closely than that of some journeyman plumber whose work he has seen for the past thirty years. He doesn't know your knowledge or ability.

If an inspection revealed a potential electrical fire hazard below the floor of your baby's bed or the fact that the wood stove flue could cause your roof to ignite, you would damned sure want to know it.

Some precautionary steps and material-usage suggestions are included in the following chapters.

The big (and sad) problem we all find with the increasingly-restrictive codes is that they demand ever more standardization, which directly curtails creativity, resulting in sameness, blandness, and mediocre design. It is sad, indeed, to consider that St. Peter's Basilica, the Taj Mahal or the Johnson Office Tower could not be built in the United States today without tremendous modifications required for code conformity! The cost of "progress" is very, very high.

Sometimes codes can be avoided by "moving out" somewhere to an area where codes are not used or enforced. Such building locations usually lack such things as public water, gas, sewer systems, and fire trucks; but you might decide that, to accomplish your dream, building around these conditions might not be a large price to pay.

If you do find your building site in the boondocks where electricity is available (practically anywhere in the country) you will most likely be subject to inspection by the company which supplies electricity to your area. This inspection will probably be less stringent than in a populated area because the electric company is mainly interested in protecting its transformers and equipment.

Remote areas sometimes have natural gas available, in which case, the same requirements might exist for gas.

States and the Environmental Protection Agency have regulations regarding private sewage systems such as septic tanks/lateral fields. Here again, you should educate yourself about what will be required and how to avoid such things as raw effluent coming to surface in your yard or running under your house (we saw this happen a few years ago at a contractor-built, custom house in the country)

and sewage getting into your drinking water source.

Often small towns and counties require that a building permit be obtained at city hall or court house even if there is no Building Regulations Department. These produce a little revenue and alert the assessor to a chance to raise some property tax. An anonymous phone call will apprise you of what local laws require.

Now let's talk about how to best deal with building codes and the people who administer and enforce them.

First off: Don't fight until every other possible alternative has been exhausted and then only fight verbally. Belligerence, causticity, and above all, an open display of your superior intelligence to that of those civil servants, will get you no place but in trouble!

Our advice to anybody building anything unconventional in an area governed by building codes is to write a letter to the Building Regulations Department of the government with immediate responsibility for that area.

State in the letter that you own, and intend to build on, a piece of property, which you should describe fully (legal description plus address). Ask the proper procedure for obtaining a building permit, what plans and specifications are required, what the front yard and side yard requirements are, and then, ask specific questions to which you must have answers to be able to continue with your planning, material-gathering, etc. You might want to ask, for instance, if pre-used lumber is allowed to be used. Is there provision in the code for post-and-beam construction, a geodesic dome, etc. Will you be allowed to install your great-grandmother's rewired chandelier?

By writing, you should get a written, signed reply and you will hopefully avoid having an in-person session with some code-book-thumping evangelist, which might prove bad for your disposition or blood pressure.

The biggest single problem common to almost all civil servants is a terrible case of negativism. It seems that all Building Regulation Department personnel think completely in terms of what is *not* allowed. This is evidenced by the fact that they will often answer "no" before they have heard your whole question.

For this reason, it is advisable to furnish the B.R.D. no more in the way of plans and specifications than they absolutely demand. The more detail you give them the more they can find to say "no" to. Even if you execute a completely detailed set of working drawings and specifications (of which you are justifiably proud and want the world to see), you would be better off presenting City Hall with a sketch on a brown paper bag showing size and shape of the structure and how far it will be from the front and side property lines. Inspectors are not anywhere near as nit-picky about things, as a general rule, as the department head, his assistant or the department engineer, because they have less book-learned ammunition.

Your best approach, regrettably, when conversing with one of these people is to recognize his or her vast knowledge about building (and life itself) and plead for help in getting poor, inexperienced, you over your hurdle. It really does often work — just don't overdo it!

After you receive a reply to your letter, you will probably know much more about how to proceed. If you have thought of other questions, or need clarification on something, it might be okay to make a phone call, but, if time permits, another letter would be best.

Remember that B.R.D. people are used to dealing with contractors and conformists and that they regard you as just some kind of kook to put up with. So, if it is suggested to you that you would be better off getting a nice home builder to work with, it is only a well-intentioned suggestion and has nothing whatever to do with what the code book or zoning manual says.

Keep in mind that how well you keep your cool through the entire project is indicative of how big a person you are. Good luck!

And remember that "putting one over on" an inspector, if it has to do with a life-safety matter, could be a deadly mistake.

■

CHAPTER FOUR

Financing

There is no magic formula which we can offer the prospective non-conventional builder for obtaining money for his or her project. We will endeavor, however, to point out a few procedures that have worked — as well as a few pitfalls to avoid.

Aside from being independently wealthy, with cash available-on-demand, the easiest house builder's financing route to follow is the conventional one of preparing plans, obtaining a bid from a "respectable house builder," going to a lending institution with land title, plans, bid and required down payment and "closing-fee." This procedure works best if the property is in a "nice" subdivision where all trees have been removed, utilities have been installed and regulations have been firmly established to assure that no one will dare display any originality or creativity — thus protecting the neighborhood and America.

This is not to imply that we are against conformity of design outright — some of our best friends are conformists — but we do urge every home owner to put some of his own, individual self on display where he lives to be better able to feel "at home" and for others to recognize as materialized personality.

As an alternative to going the house builder/bid route, it usually works to substitute yourself for the house builder or "general contractor." To do this requires that the owner/builder knows whereof he speaks! That is, of course,

vitally important whether or not one needs to get a loan.

When you prepare to build with some borrowed money, we suggest you approach the lending institutions only after you have found your building site, prepared plans, obtained bids on the portions of the work you will not be able to perform yourself and prepared itemized lists of every piece of every type material in and on the structure with your exact cost for each item. Include everything — building permit, temporary electricity, transportation costs, builder's risk insurance, utilities hook-up fees, temporary heater rental (and kerosene cost), nails, caulking, a first-aid kit — everything including a contingency amount to cover those things you won't think of and replacement of those pieces you break or miss-cut.

Money lenders are impressed with quantities of paper and figures (which constitute their whole lives, almost) if the figures and calculations seem to be well researched and valid.

Unlike the people in the Building Regulations Department, money lenders should be shown all plans, specifications and calculations, in all their splendor. Not that they will understand what they see, but they will be impressed by the effort that went into them.

If your plans are for a rather conventional structure, the banker will understand. The less conventional your presentation, the more explanation will be required. Generally speaking, the people who control and loan money are not very tuned in to matters of aesthetics or personality. But we have perceived what we think might be a change brewing in this regard. It could now be possible for a bank V.P. to be a female who lives in a stilt-house on the river and drives an Alfa Romeo.

■

Energy conservation has become big bucks and *that* bankers understand. But those people won't buy a pig-in-a-poke! Base your plans on research and gather all of the corroborating information you can find as evidence of your research. At that you will almost assuredly need to include a fossil-fuel-burning "real" heating system to suit the naturally skeptical money people. They smile at you and say, "Hey, if you build this thing and drop dead, we will own it, and we *know* a Lennox furnace will produce heat."

■

Money people (particularly those in Savings and Loan firms) realize by now that some old materials actually add to the value of a house. You will make points immediately if you tell your money person you intend to use used brick, for instance. Most have seen rail trestle beams, stained glass windows, and antique hardware used to enhance property value.

If your plans indicate the use of "barn-wood" for siding or paneling, the banker might think it will be Georgia-Pacific prefinished hard-board sheets — which he can readily accept.

Beyond that, it would be best to downplay your collection of scrounged materials even to the extent of putting reasonable new cost figures for them on your lists of itemized materials. You can think of the cost figure as "value" or "replacement cost." We doubt that any banker could comprehend a list of a hundred items which cost you nothing because general contractors gave them to you for hauling them away! Remember, if you are

loaned an amount which is more than you use, you need not take it all.

■

It is much easier to obtain a loan to build where fire protection is handy than out in the boonies. In any case, the lenders will require that you buy and maintain insurance on your property and that insurance will cost more the farther the property is from a fire-fighting company. You can be your own insurer only if you alone will be the loser if fire, wind, or water harms your property.

Wherever you build, using non-combustible materials or ones with low flame spread and high fire ratings is important. The less fire protection available to your area, the more important it becomes. Be sure to point out your good materials choices to the money people!

■

When borrowing money, the same rules apply as with dealing with building regulations people, utilities company employees, and indeed, everyone else. To derive satisfaction from the relationship, it must be a friendly and courteous one. If it becomes apparent, at some point in your conversation with a lender, that this person is not in sympathy with what you are proposing to do, it is best to ask what modifications would need to be made in your plans or scheme that would make your proposition attractive. If the guy simply doesn't like your deal under any circumstances, it will behoove you to rise, smile, shake hands, thank him for his time and wait until you are in your pickup to call him a short-sighted, ignorant, out-of-touch son-of-a-bitch. Then truck on down the street to the next money place with a smile. If your project is viable and you've got the stuff together for a good presentation, someone will buy it. Try never to act desperate or appear to be pleading for a loan. Keep in mind that lending institutions cannot survive without making loans and collecting the interest on them. Without being arrogant, know your plan is a good one, that you will be able to pay it off as prescribed and inquire what the interest rate is on a well-secured house loan like the one you need.

■

When a "home loan" is made, the lendee places not only the structure he is building up for collateral, but also the land he has bought to build it on. For this reason, as well as many others, we all need to be very honest with ourselves about what our projects will cost and what our ability will be to meet our obligations. Being overly optimistic about our job prospects, etc., or giving in to the urge to fudge a little on what our costs will be, could result in greatly reduced satisfaction with the finished product after long delays in refinancing.

■

So far we have mentioned only the conventional way of borrowing money to build, in which your property is signed over (mortgaged) to a Savings and Loan, bank or other lending institution as collateral for the money you need to build, finish, improve, or enlarge your house. This usually requires that you own your land, clear, and have at least 10 percent of the building cost — sometimes a much higher percentage.

Depending on your financial situation — how much other property you own, your income, the amount of time you will have to work on your project, etc., some other type of financing might be available.

There are always some people who need to get rid of some money and want more return than is available at the time from C.D.'s or bonds. These people know that getting more return means having less security but they often don't care too much and some enjoy risk-taking. It seldom hurts to ask a wealthy acquaintance if he, or someone he knows, has some money to loan where it will do a lot of good. That is flattering to most people.

If you can figure out how to store your materials and can stand the waiting, for instance, you might accumulate all your materials by paying cash or with a series of short-term "signature" loans.

Sometimes a person who has established a very good credit rating can borrow a substantial amount even though he does not have a great amount of assets.

One possible procedure to follow is one which will probably end up less than satisfactory — namely, starting your project before you have obtained financing or all of your materials. Unless your partially completed project shows spectacular potential (and that is highly unlikely, because most people can't visualize what you have in mind), nobody with money is likely to gamble on it developing into something of value. There is also the problem of having to find materials which will work with your plans — which is a hundred percent more difficult than adapting plans to accommodate materials.

An ideal arrangement is one called "open-end" borrowing. Ideal, that is, if the terms are fair and manageable, and you don't have to sign over your whole self, socks and soul, to the extent that you worry yourself to death and are thus cut out of your rightfully earned pleasure of accomplishment.

Besides banks and savings/loan companies, there exist many other possible money sources — each with certain advantages and disadvantages. Credit Unions are great. Mortgage companies, construction finance companies, investment clubs, etc., all have possibilities. At the time of writing, interest rates are dropping below record highs and below the maximums allowed by some states. This means that rates can vary widely from one institution (or type of institution) to the next. This also means that the "personal loan companies" or "loan sharks" will be charging much higher rates than the more legitimate lenders. We would suggest using these people's money only for very short-term loans (a few months) to get you over a hurdle or allow you to buy a hell-of-a-bargain when you can see that you can absolutely pay the loan off with money which is absolutely coming to you.

Aside from lending institutions and groups, money can be borrowed from individuals. Parents, friends, bosses, siblings, and acquaintances are all possible loan sources if they have anything to loan. But the non-monetary interest you might end up paying — in feelings of guilt, beholdenment, anger and all the other bad emotions — could well make the cost too high.

In any case, from whoever the lender, borrow in a completely business-like manner with a written, signed, dated contract stipulating exact conditions in terms of dates, quantities, interest and payment procedures! Each person must get a copy and a third should be kept by a third party.

The above cautions apply as well to obtaining a loan by means of a "co-signer." When a person co-signs a note for, and with, a person borrowing money, he is signing a legal document which states that if the other party does not, for any reason, make full payment he will be obligated, by law, to do so.

■

What are your assets? Anyone considering financing you will ask that, about the first thing, and you need to know the answer yourself to be able to plan wisely.

Some assets are obvious — like money on hand. Some are not so obvious — so an inventory should be taken. The total of your inventory list might surprise you.

Insurance policies, cars, trucks, furnishings, art and collections have appraisable value so they are assets. Plans and specifications for construction are an asset that most finance people recognize as being tangible. The more comprehensive and in-detail your plans and spec's are and the better your comprehension of them, the less head scratching and time wasting you and your hired hands will do, and the less likely you will be to mis-cut materials or build areas that will have to be torn out and

rebuilt. (Because of the psychology involved, it is usually more difficult and slower to redo something or correct a mistake than was the original task.)

Besides your savings, stocks and bonds, the next most obvious asset is your piece of real estate on which you will build. If you have made improvements on the property, such as drilling a well, installing a sewer system, creating a pond, planting trees, installing fencing, providing drives, building out-buildings etc., you have increased your assets.

Any materials you have been able to collect are assets and should be inventoried, listed and given reasonable dollar values based on what it would cost to buy them new.

■

We recommend setting up a drawing board, borrowing a T-square, scale rule and triangles and putting your project down on tracing paper (important because you will need several copies before you're finished) as thoroughly as possible, before anything else. If you don't know even the basics of how this is done, it might be time and money well spent to take a course or two in drafting. Vo-tech, high school, college, and evening courses are usually quite adequate. Business colleges sometimes have good drafting classes, but inquire about their reputation first and get your total cost quoted in advance.

You might have a friend who can help you put your ideas to paper. If you live in the vicinity of a university with architectural and/ or engineering schools, it might be possible to find a good student needing to pick up a few extra bucks. This would be good for the student and you.

Of course you can hire an architect — which is the thing to do if your project is very large and complex and you are not qualified. Residential plans do not generally need to be executed by a professional architect or engineer as do public building plans. But if you come up with a complex structure, like extended cantilevering or the like, you will need to at least hire an engineer to help you with that particular area — assuring yourself and your financier that it will stand.

■

Tools and equipment are assets whether you own them or have them available. If you can borrow your brother-in-law's backhoe, you certainly have an asset there.

There are infinite numbers of tangible assets from crystal glass and gold bars to pet pigs and antique cars. Study what you have and make a list.

You also possess gobs of intangible assets! These won't mean so much on paper, but you and the moneyman both need to be aware of them.

Things like job security and enough free time to devote to a building project are enormous assets. So are such things as good health, determination, a cooperative spouse, helpful friends and ambition.

Education and/or experience in any area of construction, mathematics, business, public relations, personnel management, etc., is probably worth more than you yourself have any idea it is. Tell the money lenders — they'll recognize it. If you have spent any time around building sites, even as a kid, you picked up lots of knowledge that you will put to good use.

On the other hand, if construction is brand new to you, you won't be hampered by traditions so much and something wonderfully innovative could develop.

Whatever your situation, we wish you the best of luck and admonish you to be persistent and determined. Knocking on doors might not be enough — it may be necessary to punch them open or cut new ones.

■

CHAPTER FIVE

Concrete (and Alternatives to Its Use)

Pouring concrete is not "building with junk and stuff." It is sometimes, however, what one must do first to have a "base" on which to assemble the "junk and stuff" he or she is able to scrounge.

It is *sometimes* what one must do. There are alternatives to building on conventional concrete footings, foundation walls and floor slabs, which will be discussed later in this chapter.

Building with concrete is generally quite simple. It can also be tricky, back-breaking and, occasionally, treacherous. Pouring concrete in subfreezing weather can create problems. Pouring in non-freezing temperatures followed by freezing can be disastrous. Rain on a freshly poured slab can wreak havoc, as it often does with freshly-dug trenches for footings, etc.

Cold combined with high humidity and/or too much water in the "mud" can necessitate staying up all night to get a slab finished. Excessive heat, on the other hand, especially combined with a stiff breeze, has caused such quick drying that concrete crews have not been able to finish everything that was poured before it was too hard to work. And heat requires sprinkling or soaking with water for a couple of days to keep the water near the surface from evaporating out and leaving a soft, crumbly face.

Under reasonable conditions, however, concrete can be poured into about any shape that can be formed to produce a hard, stable and long-lasting product.

In areas where the earth will not crumble-in, a trench for a footing or grade-beam can be dug straight, level and square-cornered, with the trench itself acting as the form. The bottom of a footing trench must be solid, virgin earth (not filled-in anywhere) with straight, clean corners between the bottom and sides.

Foundation and basement walls require forms. Forms can often be rented (set-in-place or pieces only) or they can be built out of what you might have collected. And materials used in forms can be reused later.

Concrete can be purchased "ready-mix" from your local concrete company in several different "design mixes." Just call and tell the supplier you want a quality good for whatever purpose you have — footing, foundation, floor slab, sidewalk, basement walls, driveway, etc.

You will probably have to buy a yard or yard-and-a-half from a ready-mix company to keep from paying a huge penalty, and concrete has become so expensive that the regular price is too damned high! Prepare for using enough at once to warrant the delivery. Concrete trucks usually have eight-yards capacity.

A cubic yard is twenty-seven cubic feet. A foundation wall eight inches thick, three feet tall and twenty feet long is .67 (8″ is $2/3$ of a foot) x 3 x 20, or forty cubic feet — divided by 27 — or 1½ yards of concrete.

If you want to pour concrete and save money, you can rent a portable mixer, buy portland cement, sand and aggregate and do it yourself.

It is not the purpose of this book to tell you how to do things like mix concrete. Suffice it to say that different concrete applications are best accomplished with concrete varying, to some degree, in the number of sacks of concrete per yard and the size of aggregate (chat or rocks) used. Lumber yards and libraries have books and charts.

One important fact, though, is that concrete can almost never be too dry to produce a strong, smooth product! Too much water weakens concrete. Whether you buy ready-mix or mix your own, keep it just wet enough to allow it to flow from the truck or mixer to its final resting place.

And it will help to vibrate the concrete in the forms to eliminate pockets of air. Electric vibrators can be rented if you have a lot of concrete to pour. Or you can "rod" down from the top and pound on the sides of the forms with a hammer.

Reinforcing of residential concrete is not as critical, of course, as in dam, bridge and high-rise building construction, but some is usually advisable. Wire "mesh" is manufactured in five-foot widths for floor slabs, and the standard product for this use is 6 x 6 10/10, meaning that ten gauge wire runs both ways six inches apart. You might come across almost any kind of mesh or hardware cloth that someone has abandoned which would work well as slab reinforcing.

Reinforcing bars (rebars) are manufactured in various sizes from ⅜″ diameter (#3), ½″ (#4), etc., through 2¼″ (#18). Numbers 3, 4 and 5 are common and are sometimes thrown aside at large construction sites because a few too many were ordered of various sizes, they were sent the wrong length, etc. Unless excessive, rust on rebars is not a problem, and may be beneficial.

Pieces of used galvanized and black plumbing pipes, bedrails, tubes, channels, etc., will work as reinforcing but they are hard to bend at corners and they don't tie together well because they are too smooth. Whatever you use should be bent around corners, lapped and tied (tie wire is cheap and handy for things throughout your project) and should never be allowed to touch earth or forms (allowing continuing rusting to occur).

See chapters on "Insulation, Vapor Barriers and Sealants" and "Plastics" for waterproofing and protecting slabs from "hydrostatic pressure" — water pushing up through the concrete.

■

Salvaged, old concrete is not very often usable but there are some exceptions. Concrete blocks are discussed in the chapter on "Masonry."

Other salvageable concrete products include splash blocks (at the base of downspouts — some sizeable ones are around commercial building demolition sites), concrete bumper guards from street and parking lot removal sites, and, sometimes, stepping stones. We have also seen sidewalks ripped out, leaving some nice rectangular pieces easily usable for slabs outside doors, segmented sidewalks, terracing slabs, tree "wells," etc. But they are *heavy!* Occasionally, it is possible to make a deal with a contractor or independent hauler to drop some off for you on the way to the dump for some pop money. It's worth asking.

■

All kinds of shapes are being precast out of concrete today — from the small items mentioned above to huge bridge deck beams. In any circumstance wherein things are being custom fabricated for specific purposes, mistakes will be made resulting in reject products. Reject precast concrete products must be disposed of because they occupy valuable space needed for profitable production, create living quarters for rodents, bugs and snakes, and they are depressing for management to look at.

Precast concrete railroad ties are replacing the old wood ones. Rejects make good steps, landscaping blocks, and, if you can find enough, foundation walls.

Septic tanks, burial vaults, culvert pipes, storm sewer pipes, park tables and benches, manholes, porch steps, stock tanks, feed bunkers and utility poles are all being precast in concrete, and a certain percentage of each can be obtained free-for-taking-away or at greatly reduced price because of manufacturing goof-ups or handling damage.

On the other hand, if you have no immediate need for a burial vault or a utilities manhole, then a three-and-a-half foot by seven foot workbench (the vault laid on its side on blocks) or a canned-food cellar or a two-person "scaredy-hole" might come in handy.

Concrete can be sawed with rentable equipment and considerable labor. Most cities have concrete sawing contractors now, too, but take bids first!

Most cities also have precast (prestressed) concrete manufacturers. The yellow pages will

Residence cantilevered over ravine on prestressed concrete beams. Native stone serpentine wall with mortar held to rear to appear as continuation of rock cliff.

Sometimes mismanufactured or damaged items can be repaired by an individual, with some time to spend, when repair at the factory would be unprofitable. When this is the case, the best use for a product is often the one for which it was intended.

probably list them under "Concrete-Prestressed."

The word "prestressed" comes into the picture because these big concrete beams, columns and deck sections are full of steel rods or cables which are so placed or "stretched" as to "stress" the precast piece to an engineer's specifications to be able to counter the stress put on it by other forces in the building or bridge for which it is being made. Inevitably, something will go wrong occasionally which will result in a piece not being strong enough for its intended use — but plenty strong enough to drive a car on over a ravine, hold up the earth over an underground house, cantilever a house or deck out over the side of a bluff (as shown in the photo on page 25) or become a carport floor or sidewalk.

These are often fairly flat panels. Flat, thin panels without much steel are considerably easier to cut than large prestressed structural members. Even so, if you are interested in using precast concrete pieces to build your project, it will be best to find them first and design around them.

Concrete foundation remaining after a barn was demolished.

Concrete precasters manufacture other configurations which are not, primarily, structural, so they are less packed with steel. These are often wall panels and "skirts." Sometimes these pieces are manufactured incorrectly sized or shaped and sometimes the color additives are not properly mixed or injected.

If you have not yet found your building site and you want to live in the country, you might find a place with some concrete work already done.

When an old farmhouse or barn has burned or fallen down, such as the one shown above, there sometimes remain intact foundation walls, slabs, etc., which can be taken advantage of to set the house on, incorporate into patio areas, terrace retainer walls and landscaping decoration. Be sure, however, that this stuff is strong enough to insure there will be no sinking, falling-over or crumbling.

This writer has long dreamed of finding one of those fine, big, poured-concrete silos, with enough footing, and perch a house on top, hanging over all the way around in some irregular manner. It would be particularly nice if the view included fields, woods and a meandering stream.

Broken concrete pieces are not good for much except for use in floor collector "sinks" where solar energy is stored. And even for that use, they must be well cleaned of earth and be placed in a bed of small rocks or gravel so no voids remain. (Earth is insulation and air does not store heat.)

And, speaking of insulation, remember that concrete has virtually no insulative quality at all. One inch of a commonly used insulation is equal to about forty-eight inches of concrete.

A Missouri farm residence built around a silo.
Room on top has wood casement windows for walls.

So far I haven't found one within my budget, but the photo above is of a farmhouse in southwest Missouri which incorporates a concrete silo in a different way, which is just as good and much less kooky.

The chapter on "Masonry" addresses insulation requirements, which are similar to those of concrete construction.

■

There are several ways in which frame floor construction over a crawl space is advantageous over concrete slab construction.

And now there is an alternative to concrete footing/foundation construction — where provincial building regulations don't forbid it — wood.

Actually, this alternative has been around for a while. Spotted across Norway is a group of little Thirteenth Century churches sitting on perfectly sound wood footings and foundations. The preservant recipe used for these timbers seems to have been misplaced sometime along the way, but someone recently came up with a method of pressure-treating with a preservative mixture which the manufacturers are willing to guarantee for a hundred years.

Treated lumber is discussed in more detail in the first chapter on wood, but here are a few words on the use of wood instead of concrete.

In the first place, ready-mix concrete trucks, and trucks with loads of sand and rock, are big, heavy and not easy to maneuver. This means that getting them in and out of a building spot can do severe damage to earth, rocks and plant life. Trees must sometimes be removed to allow truck access.

Lumber, on the other hand, can be carried through the trees and over the rock outcroppings with relative ease.

The most commonly used method of constructing wood foundation walls is not very unlike using concrete. A "footing" trench is dug into the earth. Chat or gravel is put in the bottom of the trench a few inches thick, and a stud wall (usually 2 x 8's on 16″ centers) is built on top of a footing. The gravel is not absolutely essential but it does allow the structure to "float" somewhat without being affected by freezing and thawing ground movement. The top of the gravel (or earth contact bed) must be level. Steps up and down in trench heights can be handled by sections of foundation walls of various heights, but each section must bear on a flat plane.

The use of a wheelbarrow might be required to place gravel in a preserved area but a small amount of gravel is easier to place this way than a much larger amount of concrete.

Another method is the use of large timbers for footings. These timbers can be new, pressure-treated material or old creosoted railroad trestle members buried into the ground to a depth, preferably, below the freeze line. A stud foundation wall is then built on top of the footings and spiked down into it.

An advantage to using large members for footings is that they will span voids below and stay in place when minor settling occurs. And on a sloping piece of ground the timber can slope with it. This requires building foundation walls which are taller on one end than the other, but that isn't difficult and will give you a chance to use any various lengths of treated lumber you have been able to find.

Nails used anywhere with exposure to the weather should be either galvanized or aluminum. In the case of nails used in treated lumber below ground, we suggest nothing but aluminum.

Stud walls used for foundations really need to be covered on the outside with exterior grade plywood. This will give the wall lateral strength (to keep the wind from blowing the house over). It will also keep the floor warmer by stopping cold air from whistling through, and it will keep out animals.

Exterior grade plywood is manufactured with glue which is waterproof. The wood veneers, however, are not waterproof, so painting of both faces and all edges before it is installed will increase life span and enhance appearance.

In a situation of an "open look" being desirable, it will be necessary to nail some "windbracing" boards onto the face of the foundation wall studs. These pieces can be placed in such a way as to improve appearance.

■

CHAPTER SIX

Masonry (Blocks, Bricks, Stone, Rocks, Marble, Adobe)

Most architectural masterpieces of the world are either totally or predominantly masonry construction. The Great Wall of China, Egyptian and Mexican pyramids, the Taj Mahal, the great Christian cathedrals, Moslem mosques and Jewish temples, the Ponti Veccio, and works by more recent architects who were artists with masonry, such as Alvar Aalto, Frank Lloyd Wright, Le Corbusier, Arne Jacobsen and Eliel and Eero Saarinen, could only be possible with masonry.

For one thing, bricks and rocks won't burn or flood away very easily, so they tend to stay around a long time.

There is one big disadvantage to building with masonry, however, and that is the very poor insulating qualities it possesses. One inch of commonly used insulation, such as fiberglass or cellulose, or 3½″ of wood, is roughly equivalent to thirty-six inches of solid stone or brick.

What this means is that the way we are building brick and stone veneer walls is wrong-side-out in terms of protection from extreme temperatures of summer and winter. Instead of acting as a barrier to heat loss, a wall or veneer of masonry will absorb and retain the ambient temperature which surrounds it. In other words, a masonry wall on the north, east or west of a building in Bismarck, North Dakota, will absorb Nature's -10° on the outside while absorbing, and quickly dissipating, whatever heat is fed to it from the inside. The wall on the south side is affected by the winter sun, while there is sunshine.

We build frame walls, insulate them and then negate some of the insulation by putting up a four-inch cold collector next to them. If we built the frame, insulated wall and put a masonry veneer on the inside, this masonry would be protected from the outside temperature and would take on the ambient inside temperature and dispel it while heat or air-conditioning is turned off, resulting in more even and comfortable temperature control.

Perfect examples of this are masonry floor "sinks" and masonry heat-collector walls (Trombe-walls) built inside the outer glass enclosure on the south side of a building. See the chapter on "Solar Energy Use" for more information.

The intention here is not to discourage building with masonry, but to encourage using it in ways that will be most beneficial.

■

Concrete blocks have become the most common masonry material used in ingredients and combinations of materials, and in lots of different sizes and shapes. The most common blocks, however, are 8″ x 8″ x 16″ nominal. "Nominal" means the actual dimensions will each be from ⅜″ to ½″ smaller than the stated size, to allow for mortar joints.

Since blocks have been used for so long a time, they are constantly available at building demolition sites, often just for the asking. If you are given some, they will need to be cleaned of their mortar. Generally this can be accomplished fairly easily with a mason's hammer or a regular carpenter's hammer and a broad cold-chisel. *Wear goggles!* And gloves can be very beneficial.

Blocks used for outside walls need insulation on the outside. Used for basement walls they need to be waterproofed and insulated against the coolness of the earth. See chapter on "Insulation, Vapor Barriers and Sealants."

There are certainly many uses for concrete blocks other than exterior wall construction, including walls around a furnace or boiler, a storm cellar, foundation walls, fireplace mass, tree "wells," basement window "wells," furnace or boiler platform, Trombe walls (see chapter on "Solar Energy Use"), garage and unheated area construction, incinerators and barbecue pits. And there are many more uses for salvaged building blocks.

When used in a heat collector wall or fireplace mass, it will help tremendously to fill the voids with mortar or sand as the blocks are being laid up.

Solid "patio" or "paver" blocks about 2" x 8" x 16" are found where block paving is being removed. Four-inch thick blocks were often used for interior partitions, and twelve-inch thick units will often be available where commercial construction is being removed.

Broken blocks and pieces can be placed in floor "sinks" for solar energy storage, but be sure the holes are filled with gravel, sand or rock and no voids are left around them.

■

Used brick became so popular several years ago that manufacturers began producing all kinds of new brick with colors (mostly green for some reason) and texture which were supposed to look "used." Real used bricks are still in demand, so they will often cost as much as, and sometimes more than, new ones. On the other hand, it is often possible to get them free-for-the-taking. The photo on page 31 shows a sidewalk of brick, before it was removed to make way for a day-care center. These bricks became a floor collector sink and vertical mass veneer in a greenhouse room. In this case the contractor gave the bricks away to keep from having to remove and dispose of them. There are two big advantages in bricks from old sidewalks, roads, and plazas: one is that there is no mortar to remove (they are imbedded in sand), and the other is that they are solid "pavers" — they don't have holes through them.

Sometimes it is possible to obtain brick which has been cleaned, free or for only a few cents apiece. And cleaning goes fairly quickly when you catch on to it. If you get hold of uncleaned brick and hire someone to clean them, make a deal with that person to pay a given amount for each whole, clean brick. Broken bricks and pieces can be thrown into floor "sinks" to collect solar energy. Keep holes and voids filled.

■

When you are scrounging around for old bricks and come across some with lettering embossed in them, grab them up! Usually they will simply show the name of the manufacturer and city. Others will have words like "Don't spit on the sidewalk." Still others might have the name of the building for which they were manufactured, special wording requested by a wealthy customer, or a contractor's name. Some special, embossed bricks went to world's fairs, etc. These bricks have special value to collectors and can be sold or traded.

Brick manufacturers sometimes have rejects, and plants, wholesalers, retailers, general contractors, masonry contractors and house builders sometimes end up with small quantities of new bricks which are not returnable or are end-of-the-run batches which can't be matched.

These small batches can often be used effectively and attractively in an owner-built house.

Sidewalk from which bricks were salvaged for use in sunroom floor.

∎

Though less common than brick, cut stone is not as much sought after. Since cut stone is usually light in color, it cannot be used effectively to gather and store heat, unless veneered with something dark or painted dark.

It is wonderful for steps, retaining walls and planters. And we have seen pieces used as table bases. If you build a fireplace where sun rays will not hit it, light-coloroed masonry is fine to use.

∎

Most areas of the country have an abundance of native stone (rocks), of one kind or another, which can be used for building. Some shapes and sizes are much easier to work with than others, and some native stone is too large to handle easily and must be broken. Round rocks are tough to use to make something both pretty and strong.

Flat stones are much easier to lay, with the flat surfaces at top and bottom (as opposed to flat surfaces exposed to view), and we feel that stacked this way a much more beautiful surface is achieved.

Often an otherwise bland or stark building can come alive with rock gardens, terrace retaining walls, posts and tree "wells." Remember that a tree will usually die if dirt is filled around its base, so tree wells are not only pretty but functional as well.

The photo on page 32 was taken in the writer's front "yard." These are various types of native Ozark Mountain rocks from creek beds, fields and glades. They cover a piece of 6-mil poly plastic (visqueen). The post was part of the parapet wall removed from the court house because it was determined to have weak mortar and be unsafe. There is no grass to mow!

∎

Rock front yard of author's house. Rocks over poly plastic to impede growth of weeds.

Marble and granite is coming out (and off) of old buildings being razed and renovated everywhere. It was used as veneer on exterior walls, corridor and lobby walls. Old bank and post office fixtures were often marble, and bank deal-plates still are. Public restroom walls and even toilet stalls were often marble, as well as many, many specialty items.

Years ago a friend in Seattle latched onto a slab about 3″ thick by 34″ x 75″ which was a table top being removed from the public morgue! It is now the floor of a small entry alcove.

And, in that same vein, a lady artist friend has several discarded marble tomb stones among scores of other "finds" in her walled-in rear garden/yard.

Marble floors and steps are common in buildings being demolished, but this stuff rarely comes out in pieces big enough to salvage.

Marble and granite must be sawed on a special wire saw, and it is doubtful you could do any good trying to cut or break it yourself. Many cities have companies which do custom cutting, and the cost is not exorbitant, but if you can figure out how to make your floor, cabinet top, table, wainscot, steps or whatever, to accommodate your stone, you'll be better off.

Entry floors, fireplace hearths, areas below wood stoves, etc., can be fashioned of broken marble, but, in our opinion, this use became "dated" several years back and lost its charm.

When stone is used for conventional flooring it should be laid in a bed of mortar not less than 2″ thick. This might require notching floor joists down and blocking floor boards between joists to support mortar bed. Use poly plastics over wood.

■

Fireplace construction requires fire bricks for lining the fire box (unless a manufactured insert is used) and clay flue tiles for safety. Sometimes these items can be found intact for reuse, but you might have to buy some.

Any time you do find firebrick, grab it quickly before it gets away. It is sort of expensive stuff and can be used for furnace linings, hearth flooring and below wood stoves.

Clay flue tiles can become fun "portholes," table bases, barbecue pits, etc.

Sometimes structural clay tile and glazed tile masonry units become available. Sizes, shapes, color, glossiness, and condition will vary so widely we can't even guess what you might find or make from what you find.

It can be fun, good stuff though, and we would like to see a picture of your masterpiece!

■

Adobe construction is a feasible approach in some areas of the country, and the "purist" with some time and a good back can manufacture his own adobe bricks of whatever size and shape he wants.

Adobe units can be sun-dried or kiln-dried. Kiln (fire) dried units do not generally require stucco on the outside, but sun-dried adobe will.

The addition of 5 to 7 percent cement to earth will make adobe units manufacturable with nearly any dirt and the cost advantage over masonry blocks is still very substantial. And dirt is fine insulation.

■

All kinds of reinforcing, wall ties, anchoring devices, etc., are available to help you avoid having cracks develop and to tie the masonry in with the rest of your structure. A visit to your local block manufacturer or brick dealer will acquaint you with what is available and there is usually someone there who can help you find what will work best for what you're doing. Take your plans with you. This person can also advise you on what mortar to use and how to make it.

One hint having to do with esthetics and stone construction: the less mortar (grout line) you allow to show, the more your rocks will look like they are laid-up "dry" and the more attractive the job will be.

And you can buy coloring to use in your mortar if you want to attain a different effect that way.

■

CHAPTER SEVEN

Wood-Framing Lumber and Large Pieces

Technology is rapidly propelling us toward a time when most of our shelters will be constructed of standardized components, mass produced, in giant factories, out of petrochemicals and alloyed metals. Standardization of products results in sameness of design of these shelters one-to-the-next.

This manufacturing standardization, coupled with increasingly more stringent building codes and regulations, is resulting in blandness and mediocrity both in commercial architecture and residential design.

Thank God for trees!

Wood is the one building material which technology cannot refine all the character out of, and it is the one which is replenishable.

Trees of amazingly varying types are growing, being harvested and, more and more, being replanted, all around the world. This chapter, however, will deal mostly with the wood from coniferous trees growing in our western states, western Canada and the yellow-pine-growing regions of our mid-south. These trees yield softwood lumber in sufficient sizes to be used for framing, timbers, utility poles, etc.

The major exceptions are these — railroad ties are hardwood (usually oak) and many old buildings being removed, for the sake of "progress" into our technologically "advancing" life, were framed with wood from the virgin growth trees standing near, or on, the construction site, whatever species they happened to be.

The barn in the photo on page 36 was constructed, almost entirely, of oak, and much of it was white oak! The studs are 2 x 4, the rafters are 2 x 6, the joists are 2 x 10, the beams are mostly 6 x 8 and the posts are 4 x 4 and 4 x 6. The decking is 1 x 4 and there are many other miscellaneous sizes, all oak.

Posts, beams and framing pieces were often cut from oak, ash, gum, poplar, birch, maple, beech, etc., depending on the area of the country. We have even found huge pieces of walnut, cherry, chestnut and birds-eye maple.

Some areas of the deep south have buildings, about to be razed, which were framed with cypress. As a matter of fact, cypress timbers can still be bought new from some small mills, on special order. Cypress has one big advantage over most other woods in that it has a high degree of resistance to rot.

When you obtain some of these large pieces of old hardwood it might be wise to consider using them for things other than framing unexposed to view.

Planing and sanding rough-sawn hardwood will bring out beauty beyond belief, but it will not be easy to find someone with the necessary equipment who will be willing to do it. The reason for this is the fact that it is virtually impossible to find all of the broken-off pieces of steel (square nails, staples, tacks, etc.) which a planer knife will hit. Someone might agree to do it if you will guarantee the cost of replacing the broken knives, but this could cost you a *lot* of money. The only way we know to really find the metal is with a metal detector (treasure-finder) and we wouldn't bet even money on that. And a custom planer isn't likely to either.

Just a few possible ways to utilize big, beautiful chunks of wood — "open" stair treads and carriages, fireplace mantel boards, table pedestals, sofa support-base-beams, spiral stair blocks and sculpture.

Long-neglected barn which yielded hundreds of board feet of oak lumber.

A good alternative is using a floor sanding machine. This will produce pretty pieces but they will not be of uniform size. And, for smoothing the edges, several pieces will need to be clamped together to provide enough surface for the sander to rest securely.

Old softwood timbers, like hardwood ones, can be spectacularly beautiful for exposed framing, as well as any number of other uses.

Wood from old buildings has had a long time to dry, and seasoned hardwood, especially oak, gets hard-like-iron. When you make your plans to use hardwood timbers, figure on using an electric drill, if at all possible. Nailing seasoned hardwood is virtually impossible without pre-drilling.

Because old lumber is so well seasoned it has also done all of the warping, bending and twisting it is going to do. For this reason it works well for gluing edge-to-edge to form blocks to carve for wall art, stair landing platforms or table tops. Edge gluing requires either the use of a jointer or a good, solid table-saw with a heavy-duty, sharp, carbide-tip saw blade (jointer blade).

■

Nineteenth century (and more recent) outmoded buildings are being taken down, or are falling down, all across our continent. Some of the best types for yielding heavy timbers and framing are obvious. Factory buildings in the northeast are being abandoned almost daily as companies move to the sunbelt or are forced out of business by more progressive foreign firms. Cotton gins, tobacco barns and warehouses are coming down in the south.

The midwest is witnessing a transformation from an agricultural base to diversification of industry — resulting in the abandonment of packing plants, agricultural warehouses, stockyard complexes, etc. The sudden skyrocketing of fuel costs have made some buildings in the mountain states impractical to keep heated well enough for use and it is often unfeasible to do major fuel-efficiency modifications of such old buildings. And abandoned mining buildings yield some of the richest finds in heavy timbers anywhere.

The west coast states seem to need to keep young, so old buildings of every type are continuously being replaced by new ones.

All across the nation there are on-going changes in our transportation systems, especially railroads. Mergers, new types of equipment, high fuel costs and decay of old facilities have caused thousands of warehouses, equipment maintenance buildings, bridges, highways, railways, offices, passenger stations, etc., to be vacated.

Population shifts from east to west, north to south, and urban to suburban have left schools, churches, auditoriums, community buildings, etc., without enough tax-base or support to keep in operation. Even minor shifts

in the American-Way-of-Life, such as new cinema buildings in shopping malls and the public's loss of interest in drive-in movies, make some good lumber (and gobs of other wonderful stuff) available to the scrounger.

Almost all of the lumber and timbers from these old buildings will be softwood cut from the great stands of coniferous trees and shipped to the factory, gin, mine or warehouse building site. Depending on the area of the country, it will most likely be fir or yellow pine, but sometimes you will find larch, spruce, hemlock, cypress, western red cedar or redwood.

In our opinion it is practically impossible for framing members to be too large, either aesthetically or from the standpoint of engineering. The major exception has to do with posts in comparison to what they are supporting. Porch columns 12″ square will look terrible holding up a roof four inches thick. On the other hand, 4″ square posts supporting a roof 12″ thick, or at least with a 12″ fascia (edge board), could be quite satisfactory.

Massive ridge, purlin and roof beams exposed to view are usually very handsome if they don't restrict head room. Deep (up and down) beams will also allow insulation to be below the deck and a ceiling installed part way down on the beam, so the beauty of an exposed frame structure is combined with sufficient insulation.

Many of these old timbers will have been milled with all sorts of shapes to create strong laps, corners, connections and bearings between posts and beams. It is not uncommon to find mortise-and-tennon, shiplap, halflap, scarf and finger joints in these timbers. When this old, hand-produced carpentry can be salvaged and used, great visual effect can be attained and value added to the property.

Often, too, dowels, wedges and splines have been used instead of nails and spikes. Bolts, nuts and washers, some of incredible size, were sometimes incorporated. Remember that they were put there with brace and bit.

If you get pieces that are too large to use, they can be ripped at a lumber mill but, like planing old lumber, hidden steel can be a problem (although not as much of one).

■

Here is a little tip about using the lingo of the lumber people with whom you might be dealing. When wood is rough-sawn, it is commonly referred to in thicknesses of "quarters." A board sawed one inch thick is a piece of 4/4 (four-quarter) lumber. One and a fourth is 5/4, 2″ is 8/4, generally up to 3″ or 12/4. Larger than that, or after it is planed to "nominal" thickness, it becomes 2 by, 4 by, or 12 by lumber or timbers.

Remember that a board foot of lumber (which is the way almost any wood you find will be priced) is 12″ x 12″ x the rough-sawn thickness. In other words, a piece of 1 x 12 four feet long will only be about ¾″ thick by 11½″ wide, but it will still be four board feet. One way to figure how many board feet are in a large timber is to multiply the thickness by the width and divide that by 12 inches to determine how many board feet are in one running (or lineal) foot of the piece. That, of course, can be multiplied by the length in feet to determine the total board footage of the piece.

Let's say you find a timber which is 7¾″ thick (you would call it 8″), 16″ wide and 22 feet long. Multiply 8″ x 16″ to get 128″. Divide the 128 by 12″ and you will know that one running foot of your timber contains 10 ²/₃ board feet. Multiply 10 ²/₃ by the length of 22 feet and you will see that one board contains an astounding 235 board feet. New price, at the time of writing, for this piece would be around a hundred and seventy-five bucks at the mill.

New lumber, except special stuff for pattern making, etc., is kiln dried in the thicknesses of 8/4 and under. A new, "green," timber will check, crack, shrink, split and, perhaps, bend and warp after you get it. A timber removed from a turn-of-the-century building will remain just like you found it.

In this way the building codes which forbid the use of used lumber are crazy, but building inspectors have neither the time or knowledge to examine each board, or batch of boards, to determine grade. New lumber is grade-stamped. Thus the law protects against the chance of under-strength lumber jeopardizing your safety.

■

Timbers from railroad trestles are different from those mentioned above in that they have been creosoted. Even though they were given this treatment many years ago, they are still not as pretty as they otherwise would be, and they emit an odor which is objectionable to many people when used indoors.

Creosote is, however, good stuff for preserving wood against the elements. For this reason, these treated timbers are just right for exterior uses such as bridge beams, deck support beams and below-floor supports which cantilever or protrude beyond the protection of the structure. They can also be used for footings where it is not feasible to pour concrete. See the "Concrete and Alternatives to its Use" chapter.

Some available railroad ties have been creosoted. Others, which are newer, have been pressure treated with more modern chemical compounds. Still others are not treated because they have been discarded as rejects before treatment. Old creosoted ties are often available where railway companies are removing whole sections of roadways and/or bridges, which is pretty common practice these days because railway companies are taxed in accordance with the total length of their systems. If a section of roadway is no longer used, it pays to tear it out and vacate the land. These ties will usually come out full-sized (about 7″ x 9″ x 8′ 3″) because the tracks are pulled up first.

Repair crews, on the other hand, replace those ties which are deemed to be so rotted-out or broken-up that they are no longer safe. This means they probably won't have too much salvageable stock left and they are often taken out by machinery which cuts them into three pieces before removal.

Reject ties can be obtained from tie mills, treatment facilities and railroad material storage yards. Concrete ties and continuous rails are beginning to replace wood ties and joints of track.

Calls to the railway companies can glean information about availability, if you have patience enough to be transferred from one person to the next with interminable sessions of being put on "hold." Your business is not, of course, very important to big railroad companies, but your tax dollars have kept them afloat with subsidies for ages, so hang right in there.

We see no enormous reason why ties could not be used in place of logs for "log cabin" construction as long as joints are sufficiently staggered and the ties are "toe-spiked" together. Solid wood is not great insulation, though, and the mortar "chinking" in the cracks is bad in terms of thermal conductivity. Roughly, 3½″ of wood and 48″ of concrete equal about one inch of fiberglass or cellulose insulation.

■

Interior of restaurant built almost entirely of salvaged materials and free native stone.

■

Railroad ties have been popular for years for use in landscaping, steps and driveway delineators. They can also serve well as light footings, support posts, and, installed close together and perpendicular to the bridge beams, the road surface for the bridge over your big ditch.

A friend and fellow scrounger, Enrique Pina, is the owner/operator of one of the world's finest eating and drinking establishments — in Portales, New Mexico. He was also general contractor, architect, materials procurer, engineer and building inspector when he built the main wing of his building. And "Henry" accomplished all of this without "benefit" of a building permit, code review or zoning department approval.

The photo on page 39 is of the interior of Henry's Los Arcos shows handmade bar-joists, native stone waterfall and some of the sixty-six utility poles which he purchased in 1981 for $36 apiece.

Besides the very long poles, shorter ones are sometimes available from utilities companies; these are what is left of poles which collided with lightning or recklessly driven automobiles.

(S4S) before the end of WWII was planed to 1⅝″ thickness and the widths were also about ⅜″ less than cut-size. Newer lumber is smaller. Two inch (8/4) softwood is now surfaced to 1 9/16″, which will be about 1½″ when dry, and widths are about ½″ less than cut-size. We mention this because it is important not to mix old and "new" pieces in the same wall or floor plan.

Hand hewn timber joinery from many years past.

Building codes, as mentioned before, sometimes prohibit the use of preused framing lumber, so check that out before you collect a lot of it. If you are not faced with having to conform to a code, or your code does not require new lumber, old framing material might save you lots of money.

Remember that old stuff has gone through the drying-out process, so "what you see is what you get."

Remember, too, that there are at least three different ways these old boards might be sized. They might have been installed "rough sawn" meaning that a 2 x 6 would actually be about 2″ x 6″. Lumber which was surfaced-four-sides

Pressure treating of lumber with chemicals to prevent rotting and termite damage has become common in the last decade.

Treated lumber is good for use where it will come into contact with earth or concrete. Even if you build with salvaged materials you might want to buy some new, treated lumber for sill plates, crawl-space posts, etc.

All wood used for foundation walls, in lieu of concrete, must be pressure treated if it is to last any time at all. Timbers can be pressure treated up to very large sizes and they are good for footings where it is not feasible to pour concrete. See the chapter on "Concrete and Alternatives to its Use."

were shredded beyond hope) could be used, sans joists, to become finished ceiling below and finished floor above.

Architect Ed Water's lake house incorporating lots of second-use building stuff.

There are other types and configurations of heavy wood members too numerous to mention, but there is one we must include to show how almost anything is possible.

Architect Ed Waters came into possession of a large amount of 4 x 6 double-tongue-and-groove, "V" joint spruce, which had been the roof deck of a restaurant/lounge building. The building filled with natural gas from an open valve and exploded early one morning, sending parts (including the roof deck) and contents all over the three-acre lot and the highway and adjoining property.

Ed found this stuff just before he started to frame his lake house (photo above). His plans included fairly conventional floor and ceiling joists, subflooring and a finish floor. Because this type of decking has considerable structural quality, the salvageable pieces (many

This application was not, however, quite as simple as it sounds. For one thing, it was learned that when wood goes through this kind of trauma it somehow loses a considerable amount of its strength. Another discovery, which required considerable expense to overcome, was that the tongues-and-grooves did not just slip together on second use. Two hundred pounds of bridge nails were sledge-hammered through the decking edge-wise to draw each piece tightly to the next. And, because there were so many chips, splits and gouges, the floor required considerable patching and filling before being sanded.

The finished product is really very nice. The decking material was used for stair treads, exterior deck and steps, opening headers, support post, railings and, to continue the design motif, rows of end-grain pieces were laid into the brick floor of the sun room.

Many other scrounged items were incorporated into this house, from an antique cut-glass window to the pre-finished roof decking panels.

■

CHAPTER EIGHT

Wood-Finish Lumber and Small Pieces

Much non-structural wood becomes what is referred to as "millwork," which we discuss in more detail in the chapter on "Millwork and Cabinets." Technically, wood paneling, trim around doors and windows, fascia boards, etc., are millwork. But in an attempt to lessen confusion regarding the reuse of old wood and the unconventional use of old and new boards, we have elected to include everything but the milled, turned and "fancy" stuff in this chapter.

Unlike the preceding chapter, this one will deal with wood of species commercially harvested around the world — and even some that are not, for one reason or another, cut for selling.

Concerning rough-sawn wood removed from old structures, the same problem exists with small pieces as with framing-sized boards, beams, etc., — resawing and planing is treacherous because hidden nails do such costly damage to saw blades and planer knives.

But old, used lumber is only one of many types of non-structural, finish pieces one can scrounge.

For instance, some old lumber has never been used — only stored. Millions of board feet of fabulous hardwood (and sometimes softwood) are resting in lofts of barns all across the continent. And millions more are in storage buildings, vacated stores, abandoned carpenter shops, residential attics, garages, etc.

This writer once bought all the lumber out of a building which had been a cabinet shop until the death of the proprietor some thirty-five years before. One side of the roof had fallen in and the city of Ash Grove, Missouri, had condemned it and requested its removal. All of the walnut, ash, cherry, pecan, chestnut, apple, etc., below the intact roof was unbelievably fine quality. Most of the boards below the cave-in had developed considerable character, to say the very least. A portion of this material was used in a kitchen remodeling project, with mirror scraps, for paneling.

In the picture on page 44, the doors at the entry to the performing-arts wing of the Springfield, Missouri Art Museum are shown with their veneer of native woods in various sizes and shapes which developed as off-falls in a custom woodworking plant.

Unlike lumber which has been nailed up somewhere, new/old stock is fairly safe to cut and plane. (Only wood from the remote forests and man-planted areas are very safe, because there is less likelihood that someone might have nailed a sign, stapled a wire, run a fence or shot a hard bullet or flintrock arrowhead into the tree.)

Sawing, planing, milling and machine-sanding can often be obtained through barter when someone has found a stored treasure. "If you will S2S (surface-two-sides) this stuff to $^{13}/_{16}''$ for me, you can keep that part of the load right there." A good barter deal can be a wonderful happening for both parties.

You won't have further outlay of money and the woodworker will spend some time that will show on his books as an expense while acquiring some material for which he has not written a check.

Some woods which were cut and stored a couple of generations back are exceedingly rare now and other species, which we get today from planted areas, came from huge, virgin-growth trees in those days. Thus you might find boards of hardwood 24″ wide and 18′ long. Chestnut, without wormholes, is not available

Art museum doors made of off-falls of various species and sizes of hardwoods laminated onto damaged solid-core doors.

Another interesting possibility with old, stored lumber is the chance of finding some rare and/or exotic stuff. A wrecking crew in Newark, New Jersey, found 307 board feet of gaboon ebony in an attic in '78. A young man in Ohio discovered that the 2,000 feet of maple that his grandfather had stored in the barn had been collected over the years at a furniture factory and was all bird's eye!

today and wormy chestnut (from dead trees) is getting holier by the minute. The last chestnut trees died many years ago.

Other trees were more plentiful in days past, and larger, so were cut more often. Red gum is a good example — just price it through regular channels today! Deep-swamp cypress, pecky cypress (the "pecks" are a tree-defect and not worm holes or rot), apple, sassafras, and persimmon are all possible finds.

And then there are some boards you might find that will absolutely play hell with tools. Hedge (often called Osage Orange) is bright yellow when first cut and mellows out to a beautiful brown. Coating hedgewood with clear lacquer will keep it bright yellow for a long time. Planers, shaper knives, hand planes, etc., will get dull just sitting next to a piece of hedge. Locust is another problem wood that way, but not nearly as bad.

◼

Beachcombers and vacationers-to-the-seaside find all kinds of wonderful boards washed up on the shore. Some of these are so exotic nobody in the country can tell what they are. If you find one of these that you can put to good use as-is, it will be fun for the memories, but it, like hedgewood, will demolish tool edges, because of all the sand and salt it has gathered. We have seen plaques and bases for artwork, stools and table-tops which were once pieces of wood lying on the beach.

◼

Strange and unknown species of wood get cut-up and fastened together into shipping crates all around the world. Sometimes, because it's about all that is available at some spot of earth, some really fine wood is used for boxes. We have seen pieces of cordia in crates from Mexico, mahoe from the Caribbean, poroba-do-campo from Brazil, and even teak from Burma. But much import crate material is cut from trees which do not produce marketable lumber because of tree size, limited quantities available or undesirability of grain, color, strength or drying characteristics.

Many items manufactured in the U.S. and Canada are still being shipped in wood boxes and crates, but this wood is usually a very low grade of a not-highly-desirable species.

Besides being in small pieces, crate lumber has one other drawback, which is not being dry. It is usually not wise to glue crate/box wood into large pieces or place it tightly side-to-side unless it doesn't matter if it checks, cracks, and opens between boards as it drys.

Patterned and collaged paneling of walls, ceilings, cabinet ends, etc., are good uses for small, irregular wood pieces.

◼

Old buildings and houses coming down contain gobs of perfectly good wood, if the need for it is established. Some of it, like boxing and sheathing, is so knotty, cracked and full of nail holes that it seems almost predestined to lead another life in the same service or burn at the city dump. Many old wood-sheathed buildings had 1 x 10's or 1 x 12's nailed onto the studs at 45° for the wind-bracing effect it offered. This put nails all over the boards, making finish-lumber salvage almost impossible.

Also, these boards tend to break up badly when being knocked off the studs. One way to salvage a whole bunch of uniform size-and-shape small wood pieces is to saw down both sides of each stud or rafter allowing the 14¼", 22¼" or whatever length pieces to fall. This does not seem to make the studs or rafters more difficult to "clean," because nails will usually follow the little pieces of sheathing boards when they are pried off. Again, these pieces can produce paneling, but we have also seen interesting screenwalls, cabinet doors, furniture and louver units made out of some of this junk that would ordinarily be hauled to the dump and burned. This is so old and dry that it can be glued together into panels or blocks and remain very stable.

◼

Removed siding, even if it can be removed without being completely ruined, is much more difficult to reuse than sheathing and boxing boards. It is not flat, S4S boards (it's been milled into drop siding, lap, shiplap, etc.); it is likely to have rotten areas; it is caked with innumerable layers of bumpy paint and it has lots of nail holes. We have seen some old siding turned inside-out and used for wall sheathing behind vertical siding where the irregularity of the surface didn't matter. We have also seen some reused inside for paneling and for members of fences, but here is one important thing to remember — much of this old material was painted in the days of leaded paint. Children can very easily chip the layers of paint off and ingest little pieces which can cause real health problems.

Salvaged interior solid-wood paneling can be a very different story from siding. It is usually milled with a major portion flat; it has varnish finish, usually, which has not built-up as thick as paint, and it has been installed with nails mainly in the milled area where they didn't show. Nails, which were used through the face of the boards should have been finish nails, which have done only minor damage.

If it can be determined which face of the old boards will be exposed to view when reused before face-nails are removed, the least damage possible need occur in nail removal. If what had been the hidden face is to be used exposed, the nails should be pounded on the point and removed as they went in. If the face side is to become the face again, the finish nails should be pulled through the piece and removed from the back side.

Much paneling and porch ceiling material was used which was milled tongue-and-groove with a bevel on each edge to produce a "V" between boards. This is called a T & G "V" Joint and is usually 1 x 4, 1 x 6, or 1 x 8. A derivative, called Car Siding, is the same thing, usually 1 x 6, except with a "V" milled down the center to give the appearance of twice as many 1 x 3's.

One widely used solid-wood paneling was 1 x 12, knotty pine, which most people thought they had to have, back in the forties and fifties. (A better promoting job to get rid of a bunch of inferior merchandise was never conceived.)

These boards were milled with a lot of detail on about 2" of one edge while the other edge had only a half of a "V" groove. Because these boards started out being about 11½" wide,

when the milled portions are all cut away, a S4S board about 9" wide can be salvaged.

The advent of "V" grooved plywood paneling, which was grooved to imitate solid-wood paneling, cheapened all "V" groove material and the photograph-finish-hardboard and other imitations of the imitation ruined, in the writer's opinion, the whole use of "V" grooved paneling.

An alternative is paneling with tongue-and-groove boards which have not been milled with more detail, such as flooring, or T & G stuff used the-other-side-out. The boards are thus differentiated more subtly by the slight irregularities of thickness and milling and the different characteristics of each piece. This often allows reuse of old, milled boards with only the area ruined by nails cut off and remilled, putting the milled side to the studs so the exposed face is smooth.

Depending on how paneling was stained, there may be streaks and spots of stain on the back side (if it was finished before being installed). Sometimes this can be removed by sanding, but a more sure way is to stain the "new" side dark.

Trim, base, door frames (jambs) and other millwork are discussed in the following chapter.

■

Hardwood and yellow pine flooring is extremely difficult (almost impossible) to salvage and sub-flooring boards are so full of nail holes that they are only good for reuse as sub-flooring or as sheathing.

Lots of miscellaneous pieces of wood can be found in and on buildings to be razed and many old structures had walls and ceilings of plaster over wood lath. Now, this stuff can really come in handy throughout any building project — particularly if you are remodeling an old place with plaster.

Every builder should have a bundle of wood shingles to use as shims and a bunch of wood lath to use for blocking, fillers, and nailing strips.

Lath also makes good tomato stakes, trellises, light duty fence pickets and layout stakes.

■

Another wood to keep in mind, wherever you are looking, is aromatic-red-cedar because it is so great for lining closets, storage boxes, etc. The cedar trees which yield this special wood are small, gnarly, and covered with little limbs so the boards will be narrow, short and knotty, but it is extremely stable, rot-resistant and workable without being dried. Even *old* boards will smell wonderful, and scare off moths, if resawn or sanded. Aromatic-red-cedar boards are often thrown away, unfortunately, because they look so ragged.

■

"Barnwood" does not, of course, have to come from a barn and it can be of almost any species, size and shape.

Most often, when found, barnwood has no paint showing, but occasionally some is visible down in the softwood depths.

Old barns and storage buildings were clad with boards of the trees found and cut around them. Northeastern barns often offer ash and maple. In the hilly Midwest, oak is more common. Southern barns and storage buildings were clad in all sorts of stuff like willow, cypress, tupelo, and gum. The plains states, where trees were not as abundant, were often built with cottonwood which didn't hold up so well. Farther west, large coniferous trees were close enough for harvesting.

For reasons stated before, it is not very feasible to plane and/or sand barnwood, but it has become popular (and there is a certain interesting beauty to it) to nail-up the weathered boards as they come down. Areas can have boards of uniform sizes or random-widths-and-lengths, installed horizontally, vertically, angled, patterned or collaged, with open joints,

boards with batten strips or whatever. Several years ago this writer found the gable-end of an old tool shed leaning against a tree where it had fallen. It was used in a lounge in Joplin, Missouri, in one piece, nailed on the wall out-of-level. Some liked it.

■

There are hundreds of little sawmills operating across the continent where an individual can take a tree to have it sliced up or where he can go to buy a bunch of pieces for his project.

Some of the mills are back in the sticks so far that OSHA hasn't found them and others OSHA can't close down because they are father-son operations with no hired help.

Depending on the indigenous trees of the area, lots of wonderful stuff comes from these little establishments, some quite reasonably priced, and some as off-falls. But lumber which has just been cut has one *big* disadvantage over lumber which has been stacked in piles for years, baked in a kiln or fastened to a building for life — it is "green," which means *wet*.

As green lumber loses its water, it shrinks and if the water leaves a board more from one side than the other (because one side is protected — like against another board), it will shrink faster on the side that dries faster — it "cups." And a board will often check and crack as it dries. Sometimes this shrinking, cupping and splitting is phenomenal.

In the early seventies, this writer specified one inch, rough-sawn, green, red oak lumber in random widths for siding on a low-budget house for an English professor and his artist wife. The vertical siding was nailed up with ring-shank nails "board-on-board" with minimum overlapping of the outer boards of one inch. Today there are several places where the black tar paper can be seen through cracks and openings between boards. This phenomenon was explained to the owners beforehand and, because of the house design, the roughness is not a problem.

If you are planning to build a few years in the future and would like to start accumulating some green lumber, you should find a place to stack it either below a roof or where you can keep it covered. Use thin strips of wood (another damned good use for salvaged wood lath) between every row of lumber, perpendicularly, to allow each piece to breathe itself dry uniformly. The cover is not as important as the drying strips so don't worry if it gets a little rain on it.

All sawmills, large and small, slice off some pieces they can't sell because of too much cracking or bending, irregular shape or highly unusual grain pattern. This writer was given such a board of walnut which was allowed to dry seven or eight years. Looking at the grain design one day the shape of a bird-in-flight sort of popped out.

Another type of sawmill off-fall consists of "slabs" which fall away from the outside of a log when the rectangular boards have all been cut from the interior. These pieces are convex on one face (sometimes with bark still hanging on) and, depending on the species, usually highly irregular in size and shape. These pieces end up in particle-board chip machines or fireplaces, but they can be used for rustic fence boards, railings, and, with a lot of work to rip them straight, as vertical or "log cabin" siding. Remember that, if used green, they will shrink — so use tar paper behind them liberally.

∎

Any place that manufactures wooden products will throw away some scraps. Custom woodworking companies and architectural millwork plants discard some nice pieces of hardwoods because it might be years before a job would come along requiring that particular specie and size. Small pieces can become furniture parts, parquet flooring, laminated tops, etc.

The Black Widow Bow Company used some exotic woods including Brazilian rosewood, in the manufacture of their championship-class archery bows back in the sixties. To be flexible yet strong enough, the rosewood pieces had to

be flawless and have a certain grain pattern. Consequently, a large percentage of the rosewood stock got thrown out. This was too much for an old scrounger to overlook and this writer's basement still has several boxes of rosewood scraps. (There would be more if his mother had not visited and, thinking it was to burn, kept herself comfy in front of a rosewood blaze in the fireplace — it burns beautifully.)

Many rings and bracelets have been made from this rosewood collection. Drill a hole the size of your sweetheart's finger (about $11/16''$ probably), bandsaw to a thickness of about $3/16''$, sand and polish out a ring or do the same thing with a hole-saw for a bracelet. Rosewood needs no finish because it is naturally oily and will polish to an unbelievably high luster just by buffing.

The next two photos on page 49 show, among other wonderful things, the pieced-together rosewood skirt, or "fascia," from the basement storehouse, on the edge of the bench/table in the studio of artist Beverly Hopkins.

∎

Carrying the scavenging of wood pieces to, perhaps, a ridiculous length, but indicating that damned near anything has some potential, the gaboon ebony sharp/flat keys from discarded old pianos make almost priceless articles such as cabinet/furniture pulls (maybe with bright chrome or brass spacer pins), inlays, jewelry and art. While they're not wood, of course, don't forget the thin pieces of ivory on the rest of the keys. Again — jewelry, mosaic work, art, etc., can be made from them.

A few years ago, it was considered almost necessary to stick with the same species of wood throughout an entire building, or, at least, area.

The pendulum has swung back toward the times of using as many species as were available in one area. Mosaic floors in castles on the Rhine and inlaid geometric or floral designs in French casework were works of some of the finest craftsmen the world ever knew.

Today, multi-specie wood furniture pieces (as well as mixtures of woods, metals, glass, plastics and ceramics) are back "in." We see this as a happy swing.

It also helps the scrounger use all the wonderful treasures which exist in such abundance for the taking.

Artist Beverly Hopkins' self-executed built-in bench. The edge is faced with thin pieces of Brazilian rosewood installed with grain running end-to-end.

Close-up of rosewood bench edge facing. The rosewood was off-falls from an archery bow manufacturer.

■

CHAPTER NINE

Millwork and Cabinets

When a building project has reached the point of being ready to "trim-out" (install trim and cabinets), it has the feeling of being "almost complete." Then, however, is when much time-consuming, painstaking and important work begins.

This is also the time when previous careful attention to the use of squares, levels and plumbs will pay high dividends.

New residential construction concepts are, thankfully, being developed and old concepts revived and improved upon. But most of these developments, including the deplorable "mobile-home" phenomenon, are based on ideas of how to circumvent the standard wood-frame (16″ in center) structure with concrete foundation, an asphalt-paper roof and a veneer of plywood, brick, plastic or aluminum horizontal siding, etc.

Quickly and/or inexpensively assembled structures, whether tin buildings, domes, caves, adobes, yurts, tepees or whatever, have certain advantages but regardless of the time and cost involved to erect the basic shelter, when it is "complete," the project has just begun.

There follows the installation of all the amenities required by modern-day people (and building-regulations), including the kitchen sink.

In other words, the "finishing touches" are about 60 percent of the project in terms of time and cost.

Millwork and cabinetry are the portions of the job that are scrutinized closely by you, the builder, and anyone else who spends time sitting around with a cup of coffee or staying overnight. We don't spend much time looking at the exterior, but we are inside a great deal.

There are an infinite number of approaches which can be taken to trimming around doors, floor-to-wall junctures, windows, etc., from no-trim-at-all (or nearly none) to massive and ornate pieces being used as the dominant architectural feature.

A "no-trim" interior, while being very sophisticated and "clean," is not the easiest and cheapest.

Standard, run-of-the-mill trim available at all lumber yards is probably the least expensive, fastest and blandest way to "trim-out."

Prefinished, imitation wood moulds made out of vinyl or hardboard is probably, in the long run, the very least expensive and most common.

While these types of trim fall within the scope of "Junk and Other Good Stuff," we feel that you would do better with "some other good stuff."

Even if you want to be quite conventional, you would probably be more pleased with fairly small trim pieces cut, square-edged, from whatever wood you really like or have been able to scrounge.

Not all trim, of course, is wood and there are no laws that we know of forbidding any material you can think of from becoming "trim."

Baseboard trim is extremely desirable to protect walls from being damaged by brooms, mops and vacuum sweepers. Trim of some sort at the bottom of windows is usually desirable for protecting the wall from wintertime water build-up on the inside of the window, which will eventually run down.

Trim at door edges, window corners, outside corners of walls where traffic will be heavy, etc., will prevent damage to walls, but plaster and sheetrock installed with metal corner-bead is quite resistant to injury too.

Doors usually need a frame or jamb-set on which to hang and a stop around one side of the frame to keep the door from swinging past center and tearing out the hinges. (Stops also serve to block sight, light, air, and noise penetration.) We say "usually" because, with much care, a door can be installed in a sheetrock, plaster or paneled opening, trimless.

Without trim around the jamb-set on both sides, or "casing," as this trim is called, it is difficult to get a good joint between wall and frame. Not impossible — only time consuming — so you might decide it would be worth it and, if so, be sure the frame is *very* securely fastened to the wall so as not to allow any movement between jamb and wall.

Trim at the juncture of walls and ceiling can serve as a way to get a good corner if sheetrock, paneling or plaster didn't turn out very well, or it can be a strong ornament. Care must be taken, however, not to use trim too large for the height of the ceiling. We have seen some horrible instances of too-heavy "cornice mould" applications. This usually comes about because of the recovery of some wonderful, big bed, cove or crown moulding (or combinations of moulding patterns) from a demolished, old, high-ceilinged and big-roomed structure and its reuse in a room with normal eight-foot ceiling height.

Not only did the eight-foot ceiling derive from standardization (with the use of 4' x 8' materials), but, importantly, to make heating systems function better. So adding height to a room to accommodate ceiling trim might be unwise.

∎

Old structures yield some fabulous millwork, but it is often so caked with layers of paint that it is difficult to even recognize it's there. Much magnificent trim, milled, and sometimes with machine or hand-carving, from maple, walnut, cherry, chestnut, etc., is hidden behind so much paint, brushed on so sloppily (and with no preparation between coats), that the detail is lost.

The new products developed for removing paint and varnish have made reclamation of old trim within reason for any dedicated scrounger. Removing finish is still a *messy* job that requires a floor area that chemicals won't ruin, in a well ventilated space. It also requires several tools, but nothing expensive. Aside from a good store-bought scraper, these instruments can be such things as various-sized screwdrivers (file them with sharp corners and edges), discarded tooth brushes, an ice pick, sandpaper, a gob of old rags and several variously shaped, hand-sized pieces of wood to use for sanding blocks.

∎

We have mentioned ceiling trim pieces. They, of course, do not need to be reused as cornice trim but can become all sorts of things, from table-edges to fireplace mantels to massive wall treatments. A word of caution, however, if you find a massive, painted cornice in a building to be razed; don't buy it until you know it is wood! Many of these old cornice installations were moulded plaster — practically impossible to salvage and reuse.

Below old cornices there was often a "picture mould." This was a detailed moulding, usually about 2″ high, with the top, back corner removed to allow pictures to be hung by means of wire and clips which hung from the top of the backed-out moulds. This stuff is great for detailed picture frames, door and window casings, furniture edges, or why not for picture moulding?

Another frequently used horizontal trim was the chair-rail. There was very little standardization of this trim and it has often been severely damaged over the years by chairs and every other conceivable object because of its vulnerable location.

■

Many old houses and buildings had large door and window trim, often 6″ wide and sometimes wider, and often with an additional piece of "L" shaped trim to the outside of the casing called "back-band." Back-band was usually without detail (it *was* the added detail) except for having rounded (eased) edges, so it works well for more modern art framing, outside corner protectors, etc. It is also great to use as the edging around plywood cabinet doors and drawer-fronts to produce a heavy yet clean, sophisticated look — "overlay" type hinges can be used by saw-notching through the back-band.

The casings, themselves, might be anything from plain, rectangular with eased edge, to very highly detailed, with flutes, beads, crowns, etc. When a lot of the same pattern of trim is available, it can become an architectural motif used throughout a project. Sometimes it can be ripped into two identical pieces of handsome trim more suitable to a smaller structure, and sometimes it can be utilized on the outside of a house if it's in a protected place.

Because these wide, often heavily detailed mouldings were difficult to miter, corners of jamb (vertical) trims and head (horizontal) trims sometimes came together at square blocks, which were thicker than the casing trims and somewhat wider. These blocks were sometimes without decoration, but were more often machine-carved with a round design of several depths. And there were some with machine or hand-carved leaves, fleur-de-lis, initials, etc.

Obviously, one who likes to decorate with ornaments and detail can go wild with these corner blocks as well as the "plinth" blocks at the bottom of the casings.

Plinth blocks are somewhat wider than the casing and a little taller than the base moulding which runs, or "dies" as carpenters say, into them. Commonly they have a beveled edge at the top and they might have any sort of design carved on the faces.

■

Baseboards of great width, sometimes made of 1 x 12's, became status symbols around the mid-nineteenth century resulting in some outlandishly wide boards being put in even modestly sized areas. Unlike cornice trim, large baseboards do not overpower smaller spaces; so, if you like the look, you can reuse baseboard, as you find it, with good results.

Baseboards were usually only moderately detailed, if at all, for practical reasons. Wide bases usually had a simple "ogee" ("S" curve) mould on top or a variation of an ogee. Consequently, old baseboards often are a good source of fine wood out of which to cut bookshelves, cabinet facings, and miscellaneous trim pieces.

The "base-shoe" or "quarter-round" mould at the juncture of the base and floor is no more than about ¾ of an inch each way and is probably not worth the bother of having to be gentle with it to save it.

■

The old casings and baseboards were milled from 4/4 stock, meaning they were from ¾″ to $^{13}/_{16}$″ thick. Plinths and corner blocks were of 5/4 stock, or heavier, so they measured 1 $^3/_{16}$″ to 1⅜″ thick. Picture moulds and some of the heavy crown, cove and bed mouldings were milled from 5/4 stock while chair rails and flat trims were mostly from 4/4.

One thing to remember about most flat trims such as bases, casings, aprons, etc., is that they were often "backed-out" to allow them to fit more tightly to irregular wall surfaces. The center portion of the piece is dadoed (hollowed) out in back.

A backed-out trim piece is fairly well ruined for use with the backside exposed.

This means that the face that is exposed to view in the old structure will need to be protected as much as possible in being taken down, in removal of nails and storage.

Trim is installed with small-headed "finish" nails which have been punched-in past the surface with a nail-set. The resulting hole has been filled and sanded smooth before finishing, or between finish coats. If these nails are simply pounded back out, as they were pounded in, the result will be considerable damage to the face.

On the other hand, if the nails are pulled on through the board so the head comes out through the back, there will be little-or-no damage except to the back, which won't show. A good claw hammer will work fine to pull nails in this way if it is used sideways, so that the nail bends and exits as the hammer rolls onto its side. Several kinds of pliers are also very handy for through-pulling finish nails.

■

There will inevitably be some damage to trim faces, and there will be new nail holes when it is reinstalled. There are some wood fillers on the market which can be applied to wood before finishing, and sanded smooth, which will actually stain about the same as the wood. The best of these that we are aware of is "Famowood" and the best source is from supply houses which furnish sandpaper and such to industry. It is necessary to use this product quickly and keep the lid on tightly when not in use to keep it from drying. It can be remoistened with alcohol, however.

The alternative to this method of filling holes in wood to be given a transparent finish is to use a colored putty applied after stain and/or one coat of finish has been applied. Good paint stores can furnish putty and color additives quite inexpensively.

The Famowood route has these advantages: (1) it will stick into very irregular holes, dents and gouges; (2) it can be sanded to a feather edge; and (3) it will not shrink. These are sometimes very important factors when working with preused millwork — depending on the degree of smoothness desired in the finished product.

Colored putty, on the other hand, is much faster to use and, again, depending on desired results, can produce a very satisfactory job.

■

We have seen some jobs accomplished with old millwork which was more crude to begin with and/or was subjected to abuses of wear or weather, where nail holes were left unfilled to fit in with the rustic nature of the project. As with any other approach to building, this one can produce a product anywhere from very pleasant to disaster.

■

Doors, windows, and transoms are covered in other sections of this book, but there are other millwork items to look for when scrounging around structures to be demolished. Probably the most exciting of these are stairways.

Most early-times, quality stair-work, whether curved, flared, or straight, was done without nails or screws. Risers were routed into treads, treads and risers were routed into skirtboards, the whole unit was placed together with dry wood wedges and the wedges were wetted. Taking squeaks out of these units required tapping the wedges in more tightly.

Newell posts (the big posts at the bottom of the stair rail and corners) and the balusters (the spindles between treads and rails) had integral dowels on the tops and bottoms which fit tightly into holes drilled into floors, treads, landings and handrails (and sometimes into ceiling trim boards when handrails terminated into ceilings of lower levels).

Over the years most stairways have been mistreated and "repaired" by people who didn't know how and couldn't afford to hire a master carpenter. Consequently nails, screws, globs of glue and chewing gum can be found holding stairways together. This will not only have marred the millwork but will make the units much more difficult to disassemble. Using extreme care in removing an old stairway will pay off many times over when reassembling, and in the appearance of your finished product.

Since old stairs will almost never work in the space allowed in a new structure, a very thorough study should be made of old and new conditions, and plans completely worked out before modification work commences.

It is *extremely* expensive to have new parts made to fit with old staircases. This is especially true of pieces of curved handrails (mock-up walls must be built to laminate the curved pieces out of ⅛″ strips), but custom turnings, routed pieces and molded trim are also often out-of-reach for the average scrounger/ builder.

∎

Larger wood mouldings (say from 2″ and up) are most commonly run on a "sticker" (with 2 cutting heads) or a "moulder" (4 heads). Smaller trim from home-town shops is often run through a shaper. To produce a particular pattern of moulds, whether a batch of two-million running feet or a single six-foot length requires a machine "set-up." A millwork company which has a machine capable of producing a trim to match a pattern from an old building would almost certainly not have the correct knife or combination of knives to match the sample. This means that knives would have to be laid-out and ground from expensive knife-steel, the knives positioned on a

"head" (or more than one), the heads attached to the machine, and aligned, before any trim could be run.

Labor cost (usually union) is very high and a set-up for a piece of special trim can involve several hours.

This is why it is important for you to be aware of this situation — suppose you were able to scrounge ten beautiful raised-panel doors with hardware, frames, stops, plinth blocks, and casing for $10 each door unit. Now, suppose that by the time you got home with your $100 treasure you were missing one three-foot long piece of head trim. It is quite possible that to replace that three-foot piece would cost another $100!

In a situation like that, it would be advisable to make a thorough search for the lost piece first and, if that failed, figure out where a plain 1 x 4 casing on one side of a door would not be objectionable, such as in a closet or bath-room.

We have seen even more exaggerated situations where someone had to have a 7½″ or an 8″ piece of some kind of trim to finish reinstalling an antique staircase or capping a nineteenth century grandfather's clock. Sometimes, in these extreme cases, a small piece can be "dinged-out" with alternating use of table-saw, jointer, chisel, router, pocket knife, and lots of elbow grease applied to sandpaper. If you undertake a project like this, *please* make a long enough piece to control on the machines and protect your fingers! If you need a curved piece and you're not an experienced mill-worker, our advice would be to go to someone who is. Hands are worth lots of dollars.

∎

Fireplace mantles and fronts, wainscot paneling, ceiling coffer trim and false-beam members are being removed from old houses across the country by the thousands. And there are often specialty items of millwork, often handcrafted by expert carpenters wanting to leave their marks in time.

Some of these are telephone niches, book-cases, fold-up ironing boards, disappearing staircases, corner cabinets, display cases and all manner of built-in beds (trundle, fold-down, drawer-into-attic, etc.).

Depending on the style of architecture you are creating, these finds can become absolute treasures, used as they were designed or modified, or they can become trading stock. A thru-the-wall phone shelf with book holder and light will certainly fetch a good trade in stainless-steel shelving for a high-tech pad, sheeting lumber or hours of some electrician's help with your wiring system. Or you could advertise "antique millwork items" and be over-run by prospective buyers.

Old commercial buildings have unbelievable millwork finds, which vary widely with the nature of the business which operated in them. Often remodeling projects over the years included hiding or covering fine millwork, either to modernize or change the building use.

At the time of writing, an old Frisco Railway freight building is being razed in Springfield, Missouri. The front, two-story section of this building was the office area. The tops of the 12″ x 12″ wood columns had very handsome, shaped "T" caps which were also 12″ x 12″ about 36″ long. Someone had very painstakingly encased these jewels with straight boards to hide the detail.

Movie theaters in downtown areas are abandoned by the thousands across the country. Only so many are needed for Little Theater groups and the like and, because of the sloped floor and high ceiling, they are extremely expensive to renovate for an alternative use, maintain, and heat/air-condition. These buildings often have ornate millwork beyond description. Here one needs to watch for grilles, doors, ticket-booth millwork and massive trim (but check it out — it might be molded plaster). Light fixtures are often fantastic and the electrically operated curtain hardware could be utilized beautifully in some solar-gain project with a photo-cell activator and insulative fabric to impede outflow of heat when there is no solar-gain.

Banks are building new and abandoning old structures to wrecking crews and remodeling contractors. Here, of course, is fine hardwood cabinetry, usually old and beat-to-hell but worth reworking. Banks and bars traditionally utilized fine woods for their cabinetry and millwork. Even rosewood and mahogany from Honduras or Africa was not uncommon. Doors, gates, check-writing desks (good for all kinds of wonderful things) and trim are old-bank salvage items as well as grilles, hardware, granite and marble.

Store buildings slated for demolition are not, usually, as filled with millwork as banks and theaters, but they are certainly worth investigating and they will sometimes surprise you. Hardware stores and "general stores" were exceptions, with miles of shelves on walls above storage cabinets — sometimes with rolling ladders for access to the high shelves. Occasionally one will find a book store with the same feature. And, of course, all old retail buildings had offices, some of which were virtually all millwork.

Office buildings of the past had much more wood trim than modern ones. For instance, there were often credenza cabinets built around the radiators below windows. And they offer gobs of doors, frames, and transoms. And, as mentioned in other chapters, some old executive offices were paneled with plywood or lumber not available today at any price.

Old classroom buildings vary greatly in millwork content and condition, but some science labs and the like in old buildings are sometimes relatively new.

"Medical Arts" buildings can yield exciting finds for millwork and cabinets. Old style dentist cabinets, optometrist cases, surgeon's instrument storage units, etc., actually had enough shallow drawers to be practical in a modern kitchen or workroom.

Why we persist in building our modern kitchen cabinets with the shallowest drawers, the 5″ ones around the tops of the lower units, is a mystery when every housewife knows how much more practical would be 2½″ ones for silverware, cutlery, and cooking utensils.

Antique porch brace reincarnated as art display shelf.

Bookshelving may have been left in professional offices, but it is more likely to have found its way to someone's garage or basement before a wrecker could buy it.

■

So far we have talked about *interior* millwork, but that is certainly not all there is! Houses and buildings falling before "progress" often yield fabulous *exterior* wooden items! Some of it has deteriorated too far to be salvageable, but much has not, even when it seems as though it certainly should have by now.

Much of the better exterior millwork was produced from swamp cypress, redwood or cedar. And white pine which was back-primed before installation and fairly well maintained could last hundreds of years.

Posts, railing, knobs, shutters, soffit mouldings, columns, pilasters, gates, entries, windows, trellises, and all sorts of Victorian and pseudo-Victorian millwork is out there ready for removal — BUT — everyone knows the supply-and-demand value, so *real* bargains (and freebies) are more and more scarce.

They are out there, however. We mentioned the freight building being razed in Springfield, Missouri. This structure has about a hundred "knee-braces" holding up the wide roof which overhangs above the docks. Made of 6″ x 6″ pieces, which have been milled and shaped, these are about 6′ front-to-back and 6′ up-and-down. The guy who "bought" the building plans to use a dozen or so — the rest he will sell very reasonably.

Artist Beverly Hopkins has used a collection of variously sized and shaped knee braces (or "corbel" or soffit braces) throughout her house. They are now art shelves, shelving braces, ceiling "braces" and wonderful ornaments. See the picture above.

So far we have talked about cabinets which can sometimes be found and salvaged. While it is often possible to scrounge cabinets which will work as incidental pieces around the house and, perhaps, vanities, it is not highly likely that enough cabinets of the same general design and size can be found for an entire kitchen.

There is nothing wrong with mixing styles of cabinets in one area if the eclectic look is your "thing," but most people would like the kitchen to be a showplace of fine cabinetry and sparkling appliances right out of *Better Homes and Gardens.*

So let's talk a little about building cabinets ourselves with materials for which we have paid as little as possible.

There are really two approaches to cabinet construction — "shop-built" (built remotely from where they will be installed), and built-in-place. Experience has shown us that, under most conditions, the "shop-built" route is the easiest and best. This is true mainly because a pile of materials cut and assembled in the middle of the floor can be "attacked" from all sides. Backs can be installed (good to impede vermin) to add strength and squareness, a cabinet can be rolled onto its back and the face sanded horizontally, etc.

This way, a cabinet should end up very straight and square. Base cabinets can even be fit with drawers and doors before installation. If "scribe area" has been provided where a cabinet fits against a wall at the end, to allow the piece to be scribed and cut to conform, exactly, to the wall and if the "chairs" (the front-to-back blocks between the bottom shelf and floor) are notched up on the bottom to compensate for some floor irregularity, installation should be quite simple. A few small wedges will be needed below the base and behind the top of the cabinet at the wall to maintain straightness and squareness when it is fastened to the wall and toenailed to the floor. Installing cabinets in an old structure will probably require more and bigger wedges.

Upper cabinets will require "scribe stiles" (one must be left loose until final installation, if the cabinet fits between two walls) and wedges also. It is advisable to install the pre-fit doors after installation.

Your library has books on cabinet building, if you need help, but what is needed most is common sense and pre-planning.

Appliance and fixture size standardization, and our having become accustomed to 36" cabinet-top height almost dictate that we follow certain norms in kitchens. All dishwashing machines, for instance, are 34½" tall to fit under a 36" high cabinet-top, 23¾" wide to fit into a 24" opening and 23½" deep to house below standard-depth tops. Most kitchen sinks are 21" front-to-back which require a top of not much less than 25½". Drop-in range tops also require these full-depth tops. Floor standing ranges are 36" tall and usually 30" wide.

There is much less standardization of size of built-in ovens, refrigerators, washers and dryers, so it is certainly advisable to know these appliance sizes before cabinets are built.

Earlier in this chapter, and in several other chapters, we have discussed how various types of materials can be incorporated into cabinetry by the scrounger/builder. One favorite way is to use pieces of solid-core doors as dividers, ends, bottoms, shelves, and even doors and drawer-fronts. (Remember to notch the top of dividers 1½" x 1½" at front and back to receive continuous 2 x 2's to be used for anchoring-to-wall and to put screws through to hold the top down.)

When whole sets of kitchen cabinets and/or vanity cabinets *are* available for removal, they often look so bad nobody wants them. What looks so awful, however, is almost always the tops, doors and drawer fronts. If you are offered a deal on some like this, check first to see if they can be unscrewed from the wall or will come out intact by prying away from wall and floor. (Built-in-place cabinets will come apart in a thousand non-reassemblable pieces.)

If you make a good deal and take home the cabinets, you can replace the top and either replace, repair or cover the doors and drawer-fronts.

Sometimes it really looks fine, for instance, to put new, natural-wood doors, or wood cover panels, over old face frames which have been painted a very contrasting color.

Slightly damaged or mismanufactured plastic-laminate (formica) tops are often available at salvage or surplus stores. Salvaged 1 x 12's are the perfect width to clean up and use for upper cabinet tops, bottoms, ends, dividers, and shelves. Scrounged ¼" plywood and paneling and hardboard makes backs and drawer bottoms. Thinner plywood, even glass-crate material, will serve perfectly well for backs. Three-fourths inch thick lumber can be utilized for cabinet faces (stiles and rails). Various moulds work to edge or decorate doors and drawer fronts. One-half inch thick lumber (even small pieces) is needed for drawer sides; all kinds of objects can become pulls and handles; "2 by" lumber is needed for 2 x 2 front and back top cleats, bases, and chairs and all manner of plywood, solid-core door stock, particle board, etc., which is in excess of ½" thick, can become ends, dividers, bottoms, shelves, and sub-tops on which to install whatever finish is desired (or available), whether formica, copper, tile, paver-bricks, plastic, glass, sheet vinyl, linoleum, parqueted wood or whatever-in-the-world.

Cabinet hardware is included in the chapter on hardware, but a few comments about hinging of doors are appropriate here because they are important to design and construction.

There are, practically speaking, three ways that cabinet doors are hinged. There are hundreds of types and styles of hinges.

First — doors can be hung flush with the cabinet face. Small, butt hinges, a continuous (piano) hinge, invisible (soss) hinges and "¾ inch wrap-around hinges" are used on this style door.

Since the door hangs inside the face frame, with the entire face of the stiles and rails exposed, these doors (and drawer fronts) must be the exact shape of their openings and slightly smaller. A uniform, small space should be left on all sides for this type installation to look good.

Plywood flush doors do not need protective strips to cover the plywood edges as much, be-

cause the edges are hidden by the face framing while the doors are closed.

Flush doors and drawer fronts are by far the most difficult to construct, adjust and maintain of the three types.

Second, doors and drawer fronts can be "inset." This is the most common cabinet door configuration. The door or drawer front is cut ⅝" larger (both ways) than the opening into which it will fit. A ⅜" x ⅜" portion is then removed from the back side all the way around which allows the piece to inset into the face frame half way while protruding out half way. If cut correctly it will accommodate hinges and will close without touching the inset edges.

The common hinge is the ⅜" inset with a face that screws into the stile (exposed to view) and with a "Z" plate which wraps around the inside of the inset and screws onto the back of the door. This hinge is easy to install and has lots of tolerance for error. There is another inset hinge which is like the common one except that the stile-mounting-plate screws into the inside edge of the stile so only the pin portion is exposed to view when the door is closed. These hinges are beasts to install. There are also some fully surface-mounted, ornamental hinges meant to resemble medieval, pounded copper or wrought iron, or hand-forged barn-door hinges — easy to install and fine if you like that sort of thing.

Third, full-overlay doors and drawer fronts. As the name implies, these doors hang, fully, on the face of the face frame. Usually the edges overlay the stiles and rails ¾" all around. An opening 16" wide and 24" high would require a door 17½" x 25½".

The distance between doors is regulated by the width of the stiles and rails. A ¾" gap (reveal) between doors and stiles would require stiles and rails to be 2¼" wide. If 1¾" thick solid-core door pieces are used for dividers, and are edged the same 1¾", the reveal between the doors will be ¼". Except for the hinge edge, however, the overlay can be varied.

The wider the gap between doors, the less noticeable are variations. Doors can be brought together almost to touching because overlay hinges have pivots at the outside edge of the door, so that the door opens within its own space (like most refrigerators). But the closer together, the more difficult the hanging and adjustment.

There are many sizes and qualities of ¾″ overlay pivot hinges. They can be installed at top and bottom of doors or can be saw-kerfed into the edge the door. If kerfed-in, all that is exposed to view is a small amount of the pivot point. New European style, "invisible" hinges are now available, but they require special installation equipment and considerable skill.

Sliding and rolling by-passing doors on cabinets are a reasonable alternative to hinged ones. You won't come up under one which has been left open, and split your head open, with sliding doors.

Plastic tracks are on the market, for bottoms and tops, for ¼″ and ¾″ thick doors. Strips of hardwood can be grooved with a table-saw using dado blades, if you want to make your own. Paraffin, such as the bars used for canning or the circles out of opened jars, will make your hand-made sliding door tracks work beautifully.

There are metal tracks, fibre glides, nylon wheels (sheaves) and ball-bearing assemblies, if you want to get elaborate. There are, also, all kinds of finger-pulls available, none of which work better than a hole about ¾″ in diameter all, or part-way, through the door.

Standard ¼″ glass can be "seamed" on edges to be wonderful sliding doors, and there are lots of other materials like plastics and hard-board perfect for this use.

Drawers are the most difficult parts of cabinets to build and install so that they work well. For someone who does not have a fully equipped cabinet shop at his disposal, the best type of drawer is a simple box (four sides and a bottom) onto which is attached a drawer front to match the style of doors used. One refinement to the simple box which is worth doing is dadoing the bottoms into the sides about a half inch up from the bottom edges. The back and front pieces can be only wide enough to touch the drawer bottom. A drawer

built this way can be more easily squared and the bottom won't come loose.

There are as many different ways to suspend drawers as people have ideas. The expenditure of considerable thought and effort here can pay off in smooth-acting, non-binding drawers that won't pull out all the way spilling contents throughout the house.

Side-mount and bottom-mount drawer glide sets, such as those made by Knapp and Vogt (KV), are available for those who want real ease of installation and really classy drawers. Though more expensive, the side-mount glides such as KV 1300 or 1395 are *far* superior to the bottom-mount ones. These have nylon wheels, with ball-bearings, which roll in steel tracks.

Hardwood slides which will work quite well can be fabricated on a table-saw with dado blades. Again, side-mounting is much superior to bottom-mount and paraffin will do wonders to produce ease and quiet.

Junk yards and scrap metal heaps are full of pieces which can be made to slide together various ways to create smooth-acting, long-lasting drawer runners.

A small block of wood screwed loosely onto the inside of the rail above the drawer will hang down into the drawer and catch it from being pulled out too far. It can be pushed up to allow the drawer to be removed.

Cabinet shelves were almost all stationary (either fit into dadoes in partitions or set on cleats) for centuries. Before the advent of panel material (plywood, particle board, etc.), fixed shelves were needed to help hold the cabinets together and keep the glued-together ends and partitions from warping.

The habit stuck with us and, to some degree, still lingers. Today, it is much easier (thus less expensive) to build, paint, transport, and install cabinets with adjustable shelves. And, of course, adjustable spaces are much more flexible and convenient.

There are several commonly applied methods of supporting shelves which adjust in height, and no end of uncommon techniques — millions of which have yet to be thought of.

Metal "standards" which apply to the inside of cabinet ends and partitions (such as KV #233 and #255) are relatively inexpensive, as are the supports (or "clips") which fit into them. Discarded store fixtures and storage units are full of these things, too.

More expensive, unless found in old fixtures or on walls being demolished, are standards which mount on the back wall and hold protruding brackets, of whatever width the shelves, which rest on them (such as KV #80 standard and #180 brackets).

Standards of all types come in various lengths, but they can be cut to any size as long as care is taken to match the little numbers on the face to assure that shelves will sit level and can be placed without going the trial-and-error route.

An adjustable-shelf system that we like requires little or no hardware and is very clean and neat in appearance. One-fourth-inch holes can be drilled about an inch to two inches apart in the ends and partitions (it's far easier to drill before assembly) in two vertical rows in place of standards.

Plastic and metal supports are available if one does not want to make his own of metal or wood dowel rod.

Wood cleats, metal strips, or even nails left protruding ½" or so, can work as shelf supports. Metal channels can be used, allowing a shelf to be pulled out some without tipping.

While cabinets and millwork can be some of your most tedious work, it can also be some of the most fun and rewarding.

■

CHAPTER TEN

Plywood, Particle Board, Fiberboard and Hardboard

Plywood is fascinating stuff — per pound it's stronger than steel and it's produced in an almost infinite number of varieties, ranging from rough-faced fir and pine (intended only for use as unexposed "sheathing"), through smooth-faced, to be painted, good-to-exotic-hardwood faced panels for cabinets, etc., to pieces manufactured in curved "moulded" sections for furniture and specialty items.

Because of its strength and relatively good stability and workability, almost any plywood is worth salvaging. You'll find something to do with scrounged plywood even if it becomes nothing more than blocking, furring strips or splice-plates.

Plywood is really not the "new" material it is often regarded as being and "veneered" (plywood) products are not necessarily "inferior" to "solid-wood" ones. Take, for example, grand pianos. Some of the very early ones had laminated (or glued-together) thin pieces of wood (veneers) which formed the unmistakable grand-piano-shaped cabinet and hinged top. This type of plywood production was going on, in very limited quantity, and at the expense of enormous numbers of man-hours, way back into the nineteenth century.

Technology advanced to such a degree that it appeared for a while that everything would one day be made of plywood. Howard Hughes even built the world's largest airplane out of it.

Then technology began to outdo itself, and now competition has almost ruined (in our

humble opinion) the plywood industry. Machines kick out sheets of rough sheeting plywood faster than it can be inspected, evidently, and much of it is trash. It is warped and crooked, the veneers are folded over and missing in large areas, the glue-lines are often hit-and-miss (allowing delamination), etc.

Hardwood faced plywoods have suffered in a different way. Before the Japanese developed methods of skimming such incredibly thin veneers from a log or timber, hardwood plywoods had enough face veneer to allow joining and sanding. Now the face skin of a piece of American walnut, Burmese teak, etc., may be as thin as 64/1000 of an inch (one can see through it before it is applied). This makes for lots of face veneers out of a tree but it also makes working plywood extremely more tricky and finished products much more susceptible to damage and less practical to repair.

And standardization, that horrible thief of originality in design, has reached the point that plywood, fiberboards, and hardboards are very difficult to find in any sizes other than 4 foot by 8 foot. There are some exceptions, but the non-standard price is so great that they are often unaffordable.

About the only panels wider than 48″ are specialty items such as ping-pong table tops (five-foot by nine-foot) and some imported panels produced in metric dimensions such as "Baltic birch" which is sometimes around five-foot in width. Flakeboard is available in many sizes — more about that below.

Lengths exceeding the magical eight-foot are somewhat more common, but again, command considerably higher prices per foot.

Plywood siding (plywood grooved in various patterns to imitate various configurations of vertical wood siding) is often available in nine-foot and ten-foot lengths. Rough sheeting plywood can be ordered in ten-foot lengths and hardwood panels can be special ordered, from some manufacturers and at exorbitantly high prices, even in twelve-foot lengths.

As mentioned before, the plywood industry in earlier times produced much more variety of widths, lengths, thicknesses, grades, species, and cores. Consequently, some buildings which are falling victim to "progress," being remodeled, etc., yield some good stuff which is otherwise impossible to obtain today. Keep alert for demolition projects which include board-rooms, cafeterias, lobbies, executive offices, etc. And old panels were not always "V" grooved. The more luxurious the space into which plywood was designed, the less likely the panels were to be an imitation of something else. The reason, as we see it, incidentally, that special lots of non-standard plywood are so difficult to obtain, and expensive if obtainable, is that some small plants used to produce them. These small plants also produced more standard plywood when a seller's market prevailed and undercut the price set by the giant producers. This was an annoyance to Georgia-Pacific, Weyerhauser, etc., so when periods of overproduction came, they produced everything anyone wanted and sold it so inexpensively that the little guys were forced to shut down. A Weyerhauser executive told this writer that this was simply an American-business-system-fact-of-life. What a sad fact-of-life for the American people to have had to swallow! Sometimes hardwood plywood was manufactured not only "book-matched" but "sequence-matched." Book matching is the technique used to assemble almost all pieces of veneer which are sliced, flat, from slabs (as opposed to "rotary cut" veneers, lathed off of round stock). Every other slice is flipped over (like a book page) to produce a continuous repeat of grain pattern across the sheet.

Sequence matching is the same process, except extended panel-by-panel with sequence numbering of the panels. This makes possible a whole room of paneling with repeated grain pattern.

We mentioned this because it is sometimes possible to salvage some sequence-matched stock. Holes have been cut for electrical and mechanical devices, but cutting strips out where needed will usually not be very noticeable in the finished product.

A problem with putting up a wall which is too blemish-free is making it look too perfect or "plastic."

Like other building products, salvaged old stuff is not the only fine plywood available for less than the going rate. Manufacturing plants produce rejects (sometimes with extremely minor problems), panels are freight damaged or mishandled at the woodworking shop or lumber yard, and some are simply surplus from jobs.

This writer once used nine sheets of ¼" Brazilian rosewood sequence-matched plywood in a study room with a seven-foot two-inch ceiling and lots of glass wall area, which required only a few inches of paneling below and above the glass. The nine sheets were 4' x 8' which had been ruined on one end by a fork-lift. The cost to the owner was 10 percent of the original price. Nine new sheets had to be re-ordered for the thirty-six by eight feet of wall in the lobby of a high-rise apartment building. One person's loss is another person's gain, and the greater the gain, the less tragic the loss.

∎

A few words about "interior" and "exterior" plywoods — exterior grade plywood consists of veneers of wood which are stuck together with "waterproof" glue. Remember that glue is not what rots — wood rots — so, while it is important to use only exterior plywood exteriorly, it is also important to seal it good with paint, varnish, or varnish, or something, and keep it sealed.

Interior grade plywood will often fall apart when subjected to moisture.

You might find pieces of a plywood called MDO (medium-density-overlay) around sign companies and new construction sites. This is great stuff for painting and shedding water. It has a thin veneer of plasticized paper on the face and is a sort of yellow-brown color.

There is also form-ply which is a product manufactured for use as concrete forms. It isn't likely that you would find any that isn't completely worn out unless you find some in a contractor's liquidation sale.

■

About any plywood can be used for one-shot concrete forms then reused, somehow, after that, if it has been given a generous coat of oil before concrete is poured against it. Form oil is available at almost any building supply dealer or, if you don't care about stained concrete, used crankcase oil will work.

■

While plywood is superior to solid-stock wood for many uses because it is in wide pieces, is smooth and has much more strength, it does, of course, have drawbacks. It does not come in long pieces, for instance. And there is always the question of what to do with edges exposed to view.

Filling, sanding, and painting is one method of treating edges, but not a very satisfactory one. Gluing and/or nailing of a strip of solid wood to the edge is usually much more practical.

For short-span shelves, or ones which will have little weight put on them, pieces of wood no wider than the thickness of the plywood can be used. Except for very light-duty shelves, plywood should be no thinner than ¾″.

For long, stationary shelves, it is always good to use wall cleats below the back, as well as the ends. A 1 x 2 wood cleat nailed into studs, or otherwise securely fastened to a wall, will allow a shelf to carry tremendous loads.

The shelf can be almost twice as strong, if the front edge has a wood piece of sufficient width. Just a 1 x 2 will do wonders to impede sagging. Wider pieces are needed for longer, wider shelves or where really heavy loads will be carried.

Adjustable shelves can have these stiffener pieces applied to both edges. If you want to see shelves sagging down from excessive load, visit your lawyer's office — better yet, his library. Why someone hasn't devised a system of shelving which will retain straightness under those hundreds of pounds of law books we can't imagine. And, ironically, the shelving in most architectural and engineering offices is about to fall (or has been propped up with product samples) under the weight of catalogs and specification books.

■

Sometimes the edges of plywood can be attractive. An architect's office in Seattle, back in the '40s, had a wall behind the reception desk which was a series of narrow rippings from ¾″ plywood applied side-by-side, vertically, with the veneer edges exposed.

Years later, this writer decided to laminate some bedside table tops 1½″ thick from ¾″ plywood rippings. They were sort of nice until they popped open about every three inches apart. The problem was probably the difference in the contraction/expansion coefficient of the plywood and the 2 x 4 pine "arm" which was glued to the bottom of the table-tops and fastened to the bottom of the bed. If screws had been used, without glue, the results might have been satisfactory.

■

One fantastic plywood product to use for building things with edges exposed, laminated and highlighted is the aforementioned "Baltic birch" plywood. This is a product produced in northern Russia and Finland, and is all birch. The uniform-thickness birch veneers, each perpendicular to the next, afford a hard, solid and attractive edge. This stuff is used for products as diverse as kitchen cabinet drawer sides, heels on women's shoes, and precision instrument mounting plates, so one might find scraps around any industrial area or cabinet plant.

■

Glass is often shipped in crates lined with a low-grade, thin (about ⅛") and flimsy plywood which protects the glass from being scratched. This stuff is close to being useless, but it is better than cardboard for some uses such as defining areas where insulation must stop to allow air to flow through an attic, etc. It can also become cabinet backs, "dust pans" and lining for air chambers such as return-air runs between joists.

■

Any time damaged, reject or discarded pieces of moulded plywood can be obtained, it should be. There is tremendous strength in curved plywood such as moulded church pews. Sections of this stuff can be used, structurally, in a zillion ways. Vertically, it can become columns and supports. Horizontally it will span good distances and support a considerable weight of people or structure.

■

Grooved plywood paneling used to be standardized at ¼" thick. Then mobile home manufacturers started using thinner stuff to save weight (and money) and paneling manufacturers "went modern" and began producing panel-

ing in metric thicknesses sort of as an excuse to put out thinner and thinner panels. Some is so thin now that the "V" grooves can be hardly deeper than paint lines. Again, what we're saying is that often the old material coming out of razed buildings is superior to new stuff.

And, too, much of the old paneling was manufactured with solid, sound backs, meaning that if the faces have been ruined beyond hope, it can be used backwards and painted or as underlayment between a rough wood floor and carpet. One-fourth inch plywood is ideal for drawer bottoms and cabinet backs, as well as ceiling material in closets, etc.

■

Relatively recently, a whole new type of building material was developed to compete with plywood. It could be manufactured out of scraps and otherwise unusable wood and sold, profitably, for considerably less money. "Flakeboard," "particle board," "chipboard," etc., while different one-from-another in some ways, are all made of little pieces of wood held together with glue. The differences have to do with sizes of wood pieces, species of woods, how densely the panels are compressed, how smoothly the faces are finished and whether or not they are manufactured in layers or "plys."

Sawing and machining these products is not easy on blades and cutters because they have such a high percentage of glue. Another drawback to working with them is the effect it can have on the human body. Formaldehyde and other ingredients in the glue tend to irritate the eyes and, sometimes, nasal passages. The possible long-term effect of formaldehyde exposure from these products has not been established at the time of writing, but we strongly recommend wearing goggles and a face mask while working with them.

There are some advantages with particle boards over plywood. It is nondirectional, meaning that it can be cut-up without regard to grain direction. It has no grain (soft and hard pattern) to "telescope" through a veneer, plastic laminate or paint applied to it, as softwood plywoods sometimes do. The higher density panels have extremely smooth faces for receiving paint.

There are some major disadvantages in these products compared to plywood, however, besides having to do with the health of people and tools. The most important has to do with strength. Plywood derives its strength from the alternating layers of wood each running perpendicular to the next. Since particle board has no grain, it doesn't help much to layer it and call it three-ply. Compared to plywood, it is certainly not a strong material.

Weight can also be a detrimental factor. Particle boards, especially the more dense ones, are considerably heavier than plywood. And, of course, no matter how much it's promoted as being nice with a clear finish, it does not have the beauty of a wood surface.

It is much harder to nail than plywood and will not hold nails or screws as well and it is more difficult to nail close to the edges without breaking.

Often it is better to scrounge some old plywood than to use new particle board, but if you run out of scrounged stuff before your project is complete, you might need to buy some particle board new, to save dollars.

Flakeboard panels are readily available in ⅜" thickness, 4' x 8', (which we do recommend for underlayment under carpet or resilient flooring if you haven't scrounged any old plywood), one-half inch x 4' x 8' and 4' x 10', ⅝" x 4' x 8' and 4' x 10' and ¾" x 25" wide x 8', 10' and 12' (very good for cabinet tops which will cover), and the same in 30", 36" and 48" widths. One and one-eighths inch x 4' x 8' is also often available. This product, unlike plywoods, fiberboards and hardboards, is priced about the same per foot for any size panels.

A word of caution about buying used, "close-out" or "bargain" particle boards of any kind. If the "bargain" has been subjected to water (a flooded storage area, leaking roof, or whatever), it is not a bargain because it was swelled. This swelling softens it and makes it extremely difficult to measure, cut, join and fit into dadoes. It's always a good idea to carry a measuring tape with you when scrounging and bargain hunting. Just measure the thickness before you pay.

Plastic-laminate (formica) covered particle board tops which are defectively manufactured or damaged, are available at many "surplus stores," "damaged freight stores," etc. These are often fantastic bargains with very slight flaws or the type of problems that a person with some ingenuity can easily remedy. These are available in 25½" widths for kitchen tops and 22" for bathroom vanities. At any rate, that is what they were manufactured to be — they offer all sorts of other possibilities.

■

Sometimes hardwood veneered particle board is available. The same holds true for it, pretty much, as for the panels mentioned above.

■

An even newer product is fiberboard-core material, which we feel is a superior product for cabinet/furniture building. It glues and holds fasteners well and is extremely stable. Of course, it is not as strong as plywood for use as shelving, but with wide edge strips and/or cleats, it can be just as good.

It reminds us of the old "upsom" or "celotex" boards except much harder, more dense, and with wood veneers.

■

Some of these "paper" boards without wood veneers, are still available new as well as salvaged.

One of our favorite building materials to buy new is a product known as "building board," "insulating wall board" or "celotex board" (because so much of it was produced by the Celotex Company for so many years).

It is usually made of ground-up cane fiber and comes in sheets ½″ thick by 4′ x 8′. Besides its relatively low cost, it has the advantages of being pre-finished white on one face, easy to cut with a utility knife, of good insulating quality (it's hard to get too much insulation), wonderful for use as light-duty tack-board (repeated thumbtacking wears it out), and it is so light weight that one person can easily handle and install it using only construction adhesive.

Obvious disadvantages are softness and low strength. It should not be used in high-traffic areas or any place where it will be bumped against by people, furniture, etc.

Pre-finished building board (celotex board) is perfect for many ceilings — it is, after all, the same product as most ceiling tiles and lay-in ceiling panels, except in full sheets. Since it cannot be taped and sanded at the joints like sheetrock, the panels should be laid out in a pleasing manner because the joints will be very visible. We have used it on a sloped (cathedral) ceiling with wood batten strips glued over the joints (or under them — however you want to think of it). We used ¼″ x 2″ S4S strips of mahogany which we cut out of a million pieces, more-or-less, of salvaged "streamline" or "oval" casing trim. (Unless you have trimmed-out an ordinary ranch-style house, you have no idea how much of this stuff there is around doors and windows and in all the places where it was the perfect stuff for closing a gap.)

If this type of detail is used with pre-planning and a degree of finesse, results can be achieved which resemble Frank Lloyd Wright detailing. Joints which run opposite to joist direction can be left sans trim or you could use, as we did, a tiny (like ⅛″ x ¾″) trim to match the larger trim, in color, or painted white to match the white board.

Suppose a person scrounged all his wall-finish materials (had no sheet-rock to finish on walls), could use a little extra ceiling insulation (and who couldn't?) and didn't like the work and extreme mess of sheetrock finishing or plastering — then celotex board could be his answer.

As with most insulating products (except the lay-in ceiling panel cousins) celotex board is all but impossible to salvage, used, because it can't be removed without breaking. Like all other products, though, it can be found as rejects, seconds and damaged goods anywhere along the line from factory to railway salvage yard to wholesale materials company to lumber yards — but new and first quality, it is one of the few good buys we know.

■

There are some similar pre-finished products that can sometimes be found at demolition sites, construction and military auctions and sales. These are the "homosote" and "tectum" type boards, two to six inches thick, used for ceiling, roofdeck and insulation all-in-one on many industrial, commercial and church buildings. Because of the strength it derives from thickness and "plying," it is often salvageable in reasonably good condition. And any of these products can be repainted — spraying before installation is recommended.

■

Other products are similar to celotex board except that they are not prefinished. Sometimes nomenclature can be a problem because these sheathing and insulating panels can be referred to as "building board" or even "celotex board." The most common of these products is the insulating sheathing board sometimes called "blackboard."

Blackboard is black because of being impregnated with asphalt for added strength, durability and protection from moisture. It is manufactured in different thicknesses — ½″ and ¾″ being the most common. Since it is intended for use as sheathing, where it will be hidden from view, it is not pretty and is extremely difficult to paint — almost impossible.

Any fiberboards, however, whether full panels or small pieces, are wonderful to cloth-cover or paste something over. Again, the insulation value is desirable, and so is the "softness" in terms of feel, sight, and sound. And a piece of cloth-covered fiberboard makes a great tack-board.

As mentioned in the "Wall and Ceiling Covering" chapter, dirt-cheap fabric, from burlap to parachute vinyl, can end up in the hands of an alert scrounger.

■

Hardboard products — often referred to by the name "masonite" because the Masonite Company made it such a commonly used material — are certainly worth watching for because of the terribly wide range of uses that they can be put to by an imaginative scrounger/builder.

While "hardboard" has been used (or misused) in some products such as trim and siding, we are going to concentrate on panel materials in this chapter.

Panels ⅛" and ¼" thick are produced and sold in two qualities — "standard grade" and "tempered."

These two grades have to do with how densely they are manufactured. More material has been compressed together to render a "tempered" board. One problem with this is that there does not seem to be a standard (as there is in particle-board manufacturing) as to how dense a panel must be to be termed "standard," "deluxe," "tempered," etc., and some tempered boards have been oil-treated in manufacture, which makes them a better exterior-use product.

There are also two types of face finishes — smooth and rough (screen back). And there are some other panels on the market such as ¹⁄₁₆" untempered board and ¼" underlayment.

"Pegboard" is perforated hardboard and is available in both ⅛" and ¼" tempered sheets. All sorts of hangers, hooks, carriers, etc., can be purchased for pegboard, to hang anything from a rag to a guitar, and it's fun to figure out new, hand-made hangers.

Wholesale and retail building materials companies sometimes have damaged products, as do freight carriers. And manufacturers sometimes run rather substantial quantities of rejects before an out-of-adjustment machine is discovered.

Salvaging panels that have been nailed and/or glued into the structure at a demolition site might not yield very large intact pieces, but down to 12" by 20", or smaller, they make drawer bottoms, stereo cabinet dividers, spice cabinet shelves and gobs of other parts.

And often hardboard panels (and cement-asbestos panels) have been installed with screws, so they can be salvaged in fairly good shape.

Discarded merchandise display units of widely diverse types, have been constructed with pegboard panels for interiors and/or exteriors. These pieces will generally come out in full pieces when the unit is disassembled. And it is amazing how many hooks and accessories have been discarded with the fixtures!

Pieces of pegboard over sewing shelves and workbenches and on walls in garages, utility rooms, closets and tool storage areas become indispensable. And framed pieces, especially in conjunction with a piece of tack board, are fun, as well as useful, anywhere from the kitchen to the teenager's bedroom — they're wonderful for costume jewelry, for instance.

Remember that pegboard must be installed with blocking between and the back and wall surface behind to allow the accessories to be attached.

Most hardboard is smooth on one side only. You might find, however, some pieces that have two slick faces — this is nice stuff for cabinet dividers between cookie sheets or record albums.

The face that is not smooth has what is called a "screen back." For most applications this face should be installed away from the use or view. The smoother the drawer bottom, of course, the easier to clean.

There are ways, however, to take advantage of a screen-back and salvaged or reject pieces will sometimes have "smooth" faces too unsightly to expose.

Several years ago we saw a department store wall surfaced with pieces of ⅛" hardboard about 12" square with a random pattern of screen-back, smooth and mirror. The hardboard used was probably masonite because it was a uniform dark brown left unfinished for years. Later it was painted an eggshell white, which was another very attractive background for merchandising, as it would have been for an area of wall or ceiling in a residence.

Spray painting screen-back allows the texture to "read-thru" while still protecting the surface and making it more cleanable.

On the other hand, artists sometimes use screen-back as a substitute for stretched canvas for oil paintings. The results are about the same with much less preparation. This writer once purchased a large unframed oil painting at an estate auction in a little town in Kansas. The auctioneer thought it was inferior because it was a "bunch of crazy globs of paint on a piece of masonite." It cost $9 and is a 1954 work of the renowned artist Katherine Kildore, done during the time of a personal crisis — very unusual, strong and, to me, beautiful. It also has some monetary value.

Another hardboard product that must be mentioned is the plastic-finished panel, which is sold by the names of marlite, abitibi board and royalcote paneling.

We are not fond of the woodgrain patterns, or the fake brick of ceramic tile, because they are so damned fake, but the smooth-faced panels, especially the solid colors, we feel, have tremendous potential.

These plastic-coated panels *do not* have the wearability of plastic-laminates, ceramic tile or even moulded fiberglass, but for inexpensive, clean and washable material above a tub or on the walls in back of the cabinet tops, it's unbeatable.

Matching plastic moulding and waterproof adhesives are available.

Hardboard also shows up on other products such as plywood and doors manufactured with hardboard faces.

■

Because it is usually less expensive to purchase sheet materials in 4' x 8' pieces than special-order for a specific need, manufacturers of such extremely far-ranging products as printed circuitry to portable storage buildings have some off-falls and rejects to dispose of. A walk through an industrial park, by the trash collection areas and over the shipping docks, will almost always reveal discards which, coupled with imagination, can become treasures. Don't be afraid to ask if something is, indeed, being thrown away because trash pickup costs the discarder money which you might save him.

Sign companies sometimes throw away pieces of MDO plywood, hardboard, particle board and, sometimes, even small pieces of colored plexiglass and other products.

Sheets of material save installation time, decrease joints to fill or cover and add stability and strength. Watch for them and use them imaginatively.

■

CHAPTER ELEVEN

Roofing Materials

Because almost all materials known to man will shed water, by using proper methods almost any material can be made into roofing. Consequently, roofing methods are mentioned in many of the chapters concerning various materials. The use of aluminum printing plates, salvaging "tin" roofing, and other metals, are addressed in the "Metals" chapter, for instance. We indicate some deterioration problems in "Plastics," mention concrete tiles in the "Concrete" chapter, discuss various uses of wood in the second chapter on "Wood," and various statements in "Insulation, Vapor Barriers and Sealants" relate directly to roofing.

The roof of a structure is, certainly, among the most important of its elements, not only because it is responsible for protecting occupants and contents from precipitation, but should also protect the rest of the structure, deflect/absorb sun rays, withstand wind and be resistant to sparks and burning debris which can drop on it. And on most types of residential design, the roof is the dominant visual expanse — thus calling for considerable attention being paid to its design, color and texture.

For these reasons, there are direct and indirect connections between this chapter and every other one in this book, as well as every aspect of your building project.

The single most important factor in determining what material to use to cover, or overlay, a roof surface is the steepness, or "pitch" of the roof plane. The more quickly water runs off of a surface the less likely it will be to penetrate that surface — so, generally speaking, the steeper the roof the less likelihood of leaks developing. And in areas where there is significant snowfall, pitching a roof surface becomes more important in terms of structure (an accumulation of wet snow can be extremely heavy) as well as water run-off (snow melts from the bottom-up and sometimes "puddles" behind dams of solid snow).

The pitch of a roof is designated by the number of vertical units of measurement per the number of horizontal units. In other words, if you draw a triangle over a roof plane with the hypotenuse following the roof slope and with the horizontal leg being twelve inches, centimeters, feet, etc., long (twelve is the number universally used for roof pitch), then the other leg will determine the pitch. A 5/12 roof rises five feet in twelve feet of horizontal run. A 12/12 roof, therefore, is sloped 45°. Thatched roofs, ski lodges, and "A" frame structures can be as steep as 24/12 or more.

Roof areas are commonly referred to in numbers of "squares" instead of square feet or square yards. A "square" is an area ten feet by ten feet square, or a hundred square feet. A structure requiring fifty squares of roofing has a roof surface of 5,000 square feet.

In determining the quantity of roofing needed for your roof surfaces, it is good to keep these things in mind:

1. Measure the roof surfaces on the elevations of your plans instead of plan view because, except for dead-flat roofs, the pitch of the roof increases surface.

2. Be sure to include areas where the roof overhangs the structure — soffits, porches, carports, etc.

3. Remember that most roofing materials need to be doubled at the starter (bottom) course.

4. Ridges require either special material or a disproportionately large amount of standard material.

5. Valleys need either special materials and/or cause considerable waste.

6. Some materials manufactured for roofing, such as asphalt or fiberglass shingles, are such that they cover a certain amount of area laid in the prescribed manner. Other products can be installed in various ways with more or less of the product "to-the-weather." The more of the product exposed (and the less covered) the less quantity needed. Pitch of roof and normal weather conditions, as well as desired appearances, will help you determine how to install your roofing.

7. Whatever material you use, and no matter how carefully you apply it, some damage is bound to occur, which should be allowed for when you calculate quantity needed.

We will concentrate on roofs with pitches of a minimum of 3/12 first.

Except for clay tiles, concrete tiles and, occasionally, metal, roofing materials are not salvageable — so we will discuss new, conventional materials (relatively inexpensive) and nonconventional roofing possibilities.

Recent years have brought significant improvements in "composition" roofing. Asphalt shingles and roll roofing became available in "seal-down" type and these shingles were made available in more and more heavy stock and in varying patterns (supposed to look like wood shingles, slate, etc.). Now fiberglass products of varying patterns, colors, fire resistances and wind protection qualities are quickly replacing asphalt products.

Many fiberglass shingles (some as light as 215 pounds per square) offer Underwriters Laboratory (UL) Class "A" fire labels and wind-resistant labels and come with a 20 year warranty. And, probably because they are thinner than asphalt, and contain less petroleum, they are becoming less expensive than asphalt.

There is a problem, however, applying fiberglass shingles when enough heat from the sun will not be available to seal them down soon. They tend to be brittle (also making them a little more difficult to install) and subject to wind damage before they are sufficiently heat-sealed together.

■

If your structure is quite small, or has only a small area with a pitched roof, it might be possible to find some "end-of-the-lot" or discontinued composition roofing at a roofing contractor's yard, wholesaler's or lumber yard. Collecting a few here and there, even of the same manufacturer, type and color, will probably be a mistake because of the huge differences in color and blend of granules from one batch, or run, to the next.

Scrounging small amounts of shingles of varying types and colors is easy, but we have never seen a roof with such a collection on it that was not hideously "cute" and ugly. It would take a fine artist, indeed, with a lot of time, to produce a decent roof with scrap roofing. If you do it, please send us a picture.

■

Sometimes roll-roofing can produce a very refined (even sophisticated) roof. It is available with the same granules and colors as shingles and a roll contains a single piece about 33 feet long. It is three feet wide and about any amount of the width can be left "to-the-weather," depending again on climatic conditions, desired visual effect and your budget.

This stuff is easy to install and maintain straight lines with uniform margins. Care must be taken at vents and flues but that is the case with any material.

It makes sense that the larger the roofing piece, the less joints and possible water routes and loose corners for wind to catch. The length can be a problem because non-uniform end joints look cheap (like a barn roof). But if your roof is designed to take all 33 foot lengths or you cut each roll into exact thirds or fourths and make a "clean" layout of your staggered joints, you can produce, in our opinion, a superior composition roof easily and economically.

■

Slate, clay and concrete tiles are terrific roofing. Expensive, heavy and difficult to install, but terrific. Another product is great and can be found in small quantities at the various roofing companies. Cement-composition shingles are manufactured in several different patterns and colors to resemble slate (they really do) and wood (they do not). The various "slate" shingles can be mixed on a roof with splendid results because the old slate roofs were often very irregular and multi-colored.

Cement-composition can be cut with a shingle cutter (which really breaks, rather than cuts); it can be borrowed or rented. Like slate, clay and concrete, cement-composition is heavy, so it requires a good structure below. It is also slow to install, so you had better do it yourself instead of paying a person union wages.

■

Wood shingles can make a very handsome roof and can be applied over spaced 1 x 4's to save decking material.

Wood shakes are also handsome, but because of their roughness and non-uniformity, they are usually applied over a tar paper roof

and serve only to look nice and protect the roof from hail.

Shingles and shakes are usually made of western-red-cedar and are manufactured in several sizes and grades. The smaller sizes and lower grades are sometimes quite affordable to buy, but they can take forever to install.

Since some terrible fires occurred in Texas in the sixties and seventies, fires which devoured rows of wood-roofed apartment buildings, many building codes have become very explicit about the need to use "fire-proofed" wood on roofs. Fireproofing costs a lot of money, so wood shingles and shakes are not being used as much as before.

As we have stated in the chapter on "Finances," and several other places throughout the book, there is nothing you can do in your building project which is more important than assuring the safety of your family. And protecting your property from fire becomes increasingly more important the farther away from a fire-fighting company you build. Aside from safety, your lending institution will be impressed by the use of fire resistant materials and their use could mean the difference between being able to obtain insurance at an affordable premium or not.

■

On roof slopes with sufficiently steep pitch, it is possible (and interesting) to use wood "siding" material of swamp cypress, redwood, or a high grade of western red cedar. This can be nailed into batts (perhaps 2 x 4's), running up the rake, over which has been applied a heavy roll-roofing. Use lead-head or rubber gasketed nails and allow the roll-roofing to conform to the shape of the roof "valleys" between the batts or "screeds." Use asphalt, rubber (cut-up innertubes) or plastic strips about a foot wide below each end joint.

We recommend this roofing application for a situation in which a bunch of good lumber has been scrounged and good horizontal lines are desired.

The use of tempered glass and plexiglass for roofing is discussed in other chapters, but two points cannot be overstated in this regard. These materials have very, very little insulating value and it is extremely difficult to waterseal around them to prevent leaking. The same things are generally true about the installation of skylights. We like the glass-roof concept, and use it often, but it is almost imperative that some sort of closable insulating system is provided. And in some climates, it is well to devise a louver system overhead to protect the interior of the structure from high, summer sun.

It is not likely, but also not impossible, that you could stumble onto a bunch of rolls or sheets of copper, brass, bronze, zinc or pewter that could be used for roof cladding. We like the appearance and watertightness of soldered joints, though they are not easy and cheap to accomplish.

Sheet metal shops can crimp edges to create "standing seams," if they would be preferable, and since aluminum is difficult to weld together, if you come by some aluminum sheet goods you could make it into standing seam roofing.

Tin roofing can sometimes be salvaged from roofs and walls of buildings being razed, as mentioned in "Metals."

Of course, new metal roofing is available at farm supply houses and lumber yards and sometimes there are sales. Know the going rate and pick some up if you find a real bargain.

Rain on a metal roof sounds nice and it takes a hell of a hail or windstorm to cause it damage.

We have never heard of reject or "seconds" roofing materials being sold by anyone, but manufacturing plants probably have some at intervals. If you are near a factory, it wouldn't hurt to inquire.

"Flat" roofs and generally those of less than 3/12 pitch are very different from those with steeper slopes. Their use could well be a separate chapter.

In the first place, we want to state that it is not true what some people state, dogmatically, about it being a terrible mistake to ever design anything with a flat roof — that it will not last — will leak no-matter-what and will be virtually impossible to patch.

Consider that over ninety percent of the industrial and commercial buildings in the United States are "flat" and, until recently, built-up roofing for flat and extremely low pitch surfaces was the only roof installation that any manufacturers would guarantee, or "bond" for any length of time. (Sometimes high-quality built-up installations can carry a thirty year bond.)

However, it is true that roofing flat, and nearly-flat, surfaces are more tricky and to achieve a quality product requires the use of good materials and craftsmanship. Since water is heavy and fluid, it runs down into any void it comes to.

It does not harm roofing membranes to be covered with water for sustained periods, and most products will last longer if they are below water constantly. Some "flat" roofs are built to hold water and some are equipped with sprinklers to maintain a water cover during hot weather for the cooling effect as well as roof protection.

But standing water and, as mentioned earlier, an accumulation of snow adds lots of weight to the building and roof structure. This weight must be considered when the structure is being designed to guard against collapse as well as "movement" of the roof plane which can do serious damage to the waterproofing membrane.

Asphalt, pitch, tar and gravel roofs are still being installed. They require hot "tar" buckets and some means to get the dangerously hot asphalt as well as the rolls of "tar paper," gravel, etc., up onto the roof. The astronomical jumps in the price of petroleum products since 1973 have drastically effected the price of built-up roofs.

Now fiberglass materials are being used, on better roofing projects, to decrease weight and labor and increase resiliency, elasticity and the life of the membrane.

We do not discourage the use of these types of roof installations, but we do discourage anyone who does not have plenty of experience and the right equipment from going the "do-it-yourself" route.

■

For small areas and some installations, a satisfactory job can be done using cold-application products — thus eliminating the need of a hot bucket — but the product is not, generally, as good, the materials are more expensive and the mess can be unbelievable.

■

Most cities have companies which spray foam on flat roofs. There is some question about the safeness of most of these products, which contain formaldehyde. We do not feel that, under normal conditions, there is reason for concern, but if a person does have real concern, he could be mighty sick, psychosomatically or otherwise — and sick is sick whatever the cause.

This spray foam has tremendous insulating quality and it stops water. It can be applied over a worn-out roof membrane and stop all leaks, but it must have a coat of black film, or aluminum — which is even better, sprayed or brushed on top to protect the foam from sun rays which can disintegrate it.

The flat roof portion of this writer's house has 1 x 6 exposed ceiling/roof decking over 3 x 10's on three-foot centers. Over the wood deck is 1½" of rigid insulation board and a pitch-and-gravel roof which didn't leak for 16 years — then the dome was nailed down. Over the old roof there is three inches of sprayed-on foam and an aluminum coating. A slotted ductboard area 16 feet square offers a sitting and sunbathing area accessible via the dome bedroom door.

■

A new "flat roof" product has come into use on commercial buildings which should soon create great possibilities for the scrounger-builder. This is the single-ply roof. A single layer of resilient plastic is spread out over an entire roof with seams cemented only where these large sheets must be joined. Properly executed, the seams are as good as the sheet, which makes an absolute barrier to water. This stuff is not usually fastened down, so it must be "ballasted" or weighted down. Various sizes of gravel and stone are used, but a ductboard (as mentioned above) would serve the same purpose.

The sheet is turned down over the roof edges and fastened.

These sheets are huge, the exact sizes depending on the manufacturer. It is not difficult to imagine all the wondrous ways this stuff could be utilized, and with what ease!

A year or so ago, we had a perfect application, but were unable to acquire the material without paying full price and a considerable lug because the area was so small. While there are several manufacturers of similar products (Carlisle seems to be the leader), there seems to be an industry-wide policy of protecting against this stuff falling into the "wrong" hands.

Since then, however, we have latched onto about 200 squares which had been down only about four years (a re-roofing project) on a Catholic church and school. A hospital needed the land for expansion — the protectionist system is already falling apart. We have several ideas about using our find, but it is stored at this time.

Single-ply material is sometimes glued-down to the deck which, probably, renders it unsalvageable. There are brand-new methods for screwing and clipping the membrane down for pitched areas or where the structure cannot bear the weight of ballast.

■

On any roof, whether flat, near-flat, moderate or steep-pitch, protrusions, joints and edges are the major problem areas in terms of application of the roofing materials and development of leaks. Edges, parapet walls, plumbing vents, flues, electrical stand-pipes, sky-lights, etc., are particularly worthy of a lot of care and attention.

But if care is exercised to get the roofing to turn up far enough (above standing snow) and adhere firmly to all surfaces, there need not be problems.

Walking on roofs does more damage than wind, hail, or rain, and often occurs at "flat" areas. We strongly suggest the use of slatted duckboards and walkboards wherever traffic is necessary or inevitable.

■

New plastic fibers have been developed, such as is now used for parachutes, which will last a long, long time when stretched out as a roof.

The need to retain warm air inside a structure in most areas of the country almost precludes fabric roofs for year-round residences, but we have seen some good uses at summer cabins, over porches, etc. It is, however, subject to vandalism, as many a convertible owner will testify.

■

CHAPTER TWELVE

Insulation, Vapor Barriers and Sealants

The glimmer of truth that finally hit us about the fact that the amount of fossil fuels on our planet is finite made the cost of what is left skyrocket a few years ago, and prices will continue almost consistently upward as we continue to use what is left.

For this reason, except for the comments dealing with safety protection, this chapter is probably the most important in the book.

We can talk all we want about proper orientation to the sun for solar energy usage, efficient heating and cooling systems, wood burning furnaces, etc., but the most important step, by far, in the building of an energy-efficient house is getting it well sealed and insulated.

First, let's talk about earth as insulation. It is fun to expound on the fact that earth is the superior insulation, but there are a few things to consider. In the central portion of our contiguous states, the temperature of the earth at, let's say, four feet below the surface is about 55° F. and the earth freezes to a depth of about 18″.

It really feels good to tour a cave in the heart of summer and experience that wonderful coolness. And if you were lost in the forest in sub-freezing cold, that same cave and interior temperature could save your life. *However*, we keep our thermostats set a little higher than 55°. The point is that, in using earth as insulation, it is important to insulate *against* the superior insulation of the earth. Building into the side of a hill calls for insulation between the walls and the earth and, to a lesser degree, between floor and earth. Building beneath the earth (with earth over the roof) requires enough dirt on top to well exceed the depth of earth freeze (frost line) in your area and insulation between the roof and the earth over it. Types and amounts of insulation will be discussed later in the chapter.

The insulating quality of a material is most commonly spoken of in "R" value. The R stands for resistance, which means the resistance of the material to the penetration of cold or heat. For instance, 3½″-thick fiberglass batt insulation has an R value of 11; six-inch batts of the same product are rated R-19.

Just about any material has some R value. Some is surprisingly high and some unbelievably low (metals are negligible).

As a rule of thumb, an inch of commonly used building insulation is equal in R value to 48″ of concrete, 36″ of brick or 3½″ of wood.

Generally speaking, the insulating quality of a material is provided by the pockets of air trapped in the material. For this reason some very unconventional materials, properly used, can insulate very well.

And here is something to remember. Insulation material will resist the pass-through of heat or cold only where it is in-place. Examples: A wall built with 2 x 4's on 16″ centers and insulated with 3½″ batts will have an insulation value of R-11 *only* where the insulation is (the studs between offer less than one-third of that amount). Rigid insulating panels over a flat roof may have an R value of 6 and be nailed down to the wood deck with so many nails that the insulating effect is practically negated because the steel nails conduct heat so well. A 30'-long exterior wall with R-19 insulation has three windows four feet in width, rendering an average resistance to heat loss of R-8.

So! Take care to install your insulation (whatever stuff you find for the purpose) as continuously as possible, use tightly sealed doors and windows (with insulating shutters, etc., if possible) and don't penetrate the insulation with anything metallic (use adhesives).

Commonly used insulation materials for ceilings and wall cavities (between studs) are fiberglass (the stuff that looks like cotton candy), rock wool (which looks like it sounds) and cellulose (which looks like a bunch of little pieces of newspaper all stuck together, because that's what it is).

There are many, many other types of insulating materials manufactured for that purpose and many, many others intended for other uses. Among the latter are sprays, foams, polystyrene (styrofoam) panels, foil-encased urethane foam, pipe covers, rigid roof panels, wood-fiber boards, "blackboard" sheathing panels, cork, "homosote," perlite, vermiculite, duct lining (inside and outside), acoustical ceiling panels, etc., etc.

Some products have vapor barriers (such as the craft paper on one side of some insulation batts); some do not. Some have reflective faces on one or both sides (such as urethane board).

■

For some strange reason, quantities of new insulation seem to be sitting around collecting dust in all sorts of weird places. Consequently, it can be found at garage sales, all kinds of auctions, rummage sales, etc. Keep an eye out for partial rolls of batts, broken bags of blanket stuff and scraps.

Pieces of styrofoam and bead-board are constantly being wind-blown from construction sites and off trucks.

■

Sheet metal contractors (heating and air-conditioning companies) throw away pieces of duct lining board when it is cut to fit ducts and there are off-falls. These pieces are sometimes quite small, but when pieced together, especially when more than one layer is used and the joints are staggered, extremely good heat-loss resistance can be attained (about R-4 per inch!).

Acoustical ceiling contractors, like everyone else, damage a certain amount of "lay-in" panels and ceiling tiles. They also accumulate some factory rejects and discontinued patterns. This material can be spray-painted and used as it was designed to be or for walls in some areas where there will not be traffic (it's great to pin posters to, etc.).

Of course, there's no law that says anything has to be used as a finish material, and acoustical tiles and panels have considerable insulating value when used to fill cavities, placed between floor joists, etc.

With a phone call to the acoustical ceiling company, you can find out if anything is available. There are numerous factories around the country producing this stuff, and some rejects are inevitable. If there's a plant near you, why not check it out?

■

It would not be practical, ordinarily, to salvage used insulation. Insulation which has been blown into attics and walls is filthy dirty, unhealthy to breathe and difficult to gather and package. Batts get torn to shreds and fiberglass (even new) sticks into you, scratches you and fills your clothes with tiny slivers of glass. Panels of rigid boards can only be removed in shreds.

We are not saying you can't (with time and effort) salvage insulation, but the ratio of new-material cost to time spent and aggravation might add up to a poor use of your time.

An exception to this is styrofoam boards and blocks you might find at the site of a cold-storage building or ice plant being razed. Thick polystyrene can be easily cut into thinner slices or filler blocks.

■

All sorts of fragile and finished merchandise is packed in all sorts of packing material, from balls of newspaper to carpet scraps to *styrofoam!* Some is molded into shape for merchandise to fit into and some is in the form of little balls or pellets. The formed pieces can be cut into uniform sizes and shapes and pieced together. The little blobs will make the best imaginable ceiling "blanket" or cavity filler. It will take a lot of it, but there's a lot of it out there. Find someone who receives lots of this stuff and will put it aside for you.

When you use any of this stuff, don't forget the vapor barrier.

■

There are many naturally good insulating materials which, because of their nature, can only be used under certain conditions.

Paper (especially newspaper and corrugated cardboard) is exceptionally fine insulation, as is hay, straw and pine boughs. These materials have two almost insurmountable drawbacks. They are extremely easy to ignite and burn, and they attract insects and rodents.

Some farmers pile bales of straw up against the north wall of their houses. They keep a lot warmer using much less fuel, and they have a zoo of rats, field mice, snakes and birds right there close.

Cellulose insulation is, as mentioned above, made of newspaper pieces and, years ago,

some manufacturers were producing stuff that was not properly treated for insect or flame retardation. Regulations of the industry put a stop to these bad products and some recent tests have shown cellulose to be superior to rock-wool and fiberglass for flame spread quality.

■

"Insulating glass" is a name given to glass panes which consist of two or three layers of glass hermetically sealed onto a band around the perimeter, with space between the glass pieces. The name often misleads people. Five-eighths inch "insulating glass" has an R value of about 1½. That isn't much between you and the invading polar air mass, but it beats a single layer of glass, which is less than R-1.

We strongly recommend the use of some insulating panels or closable shutters for all windows. There are, also, new insulating fabrics that are effective.

Sometimes it makes sense to cut insulating panels to fit windows on the north side of the house and leave them there all winter. East and west walls are not quite as critical. You need to have south-facing windows clear of obstructions to allow sunrays to come in and produce warmth during hours of sunshine but insulatable at night and when it's cloudy so you can trap your free heat inside.

Some insulations, such as polystyrenes, will deteriorate when subjected to direct sunlight, so should be covered for use as window panels. Urethane, with foil on both faces, is wonderful for this use because the foil reflects interior heat back in the winter and the sun back out in the summer. This summer sun reflection might cause blinding problems in the neighborhood, however.

What we have said about insulating windows is doubly true for glass roof areas in greenhouses and sun rooms, because heat rises and tries hardest to escape upward.

■

An otherwise well insulated structure can still be a problem to heat and be comfortable in if attention is not paid to sealing cracks and joints between adjacent materials, around doors and windows, junctures of walls to roofs, slabs and foundations, etc., to prevent air infiltration. We want to emphasize this because, when a house is built of whatever-one-can-find, it is likely to have more than normal irregularities.

Gaps can be stuffed with scrap insulation, but don't stuff too tightly, because compacted insulation loses its R value.

Tighter joints, cracks and corners (and behind exterior trim) should be caulked.

■

Vapor barriers and moisture protection films are another important aspect of any building project. A vapor barrier is a thin membrane of some material which will not allow moisture to pass through. It is placed between the heated/air-conditioned space and the insulation (behind the ceiling or wall finish) to stop vaporized (wet) air from forming where hot and cold meet, causing staining and eventual rotting.

Polyethylene plastic sheeting (visqueen), about four or six mil thickness, is really good, easy to install and inexpensive vapor barrier material. It is quite likely that you will have some left over from what you bought to protect your materials from the elements in transit and storage.

Other materials, such as tar paper, can be used, and you might have some of it left over from earlier work.

A mechanical stapler is about the best tool we know for installing vapor barriers and one will come in very handy all through the job.

■

About half of the moisture in a house which has a crawl-space between the earth and the floor comes up from earth below the house. A layer of material like the vapor barrier stuff can make your house more comfortable and protect it and your clothes from mildew in the hot season.

■

Crawl spaces and attics need to breathe in the hot season, to prevent rot below the house and excessive heat above. We have seen all sorts of interesting closable vents and louvers made out of scraps. Don't forget to install screen wire in them.

■

Weatherstripping of doors is yet another important step in building an energy-efficient house. There are jillions of different types, and most are inexpensive, but you will find all sorts of thin pieces of metal, nylon, foam and rubber which you can make work as well as new stuff.

■

CHAPTER THIRTEEN

Metals

One of the major differences in commercial and residential construction is in the much more common use of metals in commercial buildings. Commercial building contractors tend to have more equipment for using metals than house builders need. A contractor who builds steel-frame buildings will have cutting and welding equipment, heavy drills, etc., as well as equipment for moving and lifting heavy pieces.

This does not mean, however, that steel structural members, copper roofing or aluminum extrusions should not be used in houses.

Pipe posts can be used very effectively in basements, crawl spaces and walls. Steel beams and open-web joists (bar joists) can be smaller than wood members, thus allowing more head room in basements, for instance. And steel members of all kinds (pipes, square tubes, "H" beams, angles, channels, etc.) can often be purchased at, or below, scrap iron price at building and bridge demolition sites and junk yards. And you can often get it torch-cut very inexpensively before you move it, if there is still equipment at the job-site. Simple end plates, caps and pieces can be easily welded on with the same equipment. Hauling and placing heavy pieces might present problems, though. Determine weight and figure a plan of moving and installing before you jump into a deal that will cause you a back injury or that you might have to give up on.

We recently purchased, cheap, three steel girder trusses sixty feet long, six feet high at one end and four feet, eleven inches high at the opposite end. These we set and pinned into trenches in a solid limestone ledge. (The trenches were filled with concrete for grouting and stabilization.)

Bar joists are reasonably light-weight and have tremendous strength. They can also be cut to length with a hacksaw, if you know where to cut. But find an engineer to advise you about any steel members, if you don't know exactly what you have, the amount of load you will be subjecting it to, etc. It is easy to draw or end-trace a profile of a steel beam, channel or other piece to take to an engineer. If you do have a basic knowledge of loads, sizes and configurations, books and charts are available at the library to use to determine safe lengths of span, spacing, etc.

Holes for fastening-in-place and joining members can be drilled fairly easily with an electric drill (preferably one with variable speed), a good steel bit and a center-punch (to make an indentation for the bit to start in, to keep it from skittering all over the piece). Cutting oil or plain-old lubricating oil will speed your job and save your bit on heavier pieces and when you are doing lots of drilling at the same time.

■

Metal roofing is sometimes available where it is being removed. It is of many different types and materials.

Old barns yield corrugated galvanized steel, "tin" panels and, sometimes, flat panels with "standing seams" which allow one piece to lap the next for more-or-less watertight joints.

In good shape, this can be reused as roofing on roof planes which are steep enough to shed rain quickly and not allow a thick accumulation of snow (which can melt, fill the laps of corrugations or standing seams, and drip inside the house). The sound of rain on a tin roof is very pleasant to most people.

There are also other good uses for this stuff. It is great, painted black, to use in solar collector panels, for instance. You can find literature on designing and building these at your neighborhood college science department or public library. Corrugated panels have considerable strength, so they can be used to retain earth, form concrete or, with hoops, to circle aboveground pools.

"Metal buildings" yield steel panels from walls as well as roofs and the old "Quonset Hut" buildings have interesting curved pieces.

Then there is the possibility of finding old copper roofing, guttering, scupper boxes and downspouts! If you do find some, grab it fast and figure out what to do with it later. If you don't use it for its intended purpose, there are zillions of other good things to do with it, from cutting out street address numbers, to covering furniture and cabinet tops, to forming summer sun shields over south-wall windows. (Frank Lloyd Wright made extensive use of sheet copper and allowed it to gain character over the years as it "patinaed" and turned shades of green.) And, as always, it can be sold or traded at the junk yard.

Other varieties of metal roofing might fall your way, any of which could be used in solar collectors or any number of other imaginative ways.

Grates and grilles can be utilized effectively for heating, air-conditioning and ventilating systems, and some are ornately cast out of iron, zinc, brass, etc. Old tanks, pipes, pipe fittings, radiators, valves, coils, etc., can be used in your hand-made mechanical system, and pipes are good for supports, furniture legs, railings, bed posts, lamp stems, closet rods, fences, etc., etc.

Some printing companies use aluminum sheets for printing plates. Once used on both faces they are disposed of, and so can be obtained cheaply. This is very thin material for which we have found several good uses. It makes good termite shielding between concrete foundations and wood; it works well for flashing; it can be stapled or tacked over knot holes to keep mice out; and, as the photo of the dome (on page 83) shows, it can be used for shingles. When it is used for roofing or siding it needs to be used liberally. Start off with two or three thicknesses at the bottom and continue up with only three or four inches exposed to-the-weather. Roofing mastic is cheap and it can help to prevent wind damage. We find that a hand stapler with ⅜" or ½" staples is the fastest and easiest installation tool. Before painting aluminum or galvanized metal it is good to wash it with vinegar to make the paint stick better.

Junk yards can be paradise islands for people scrounging for stuff to build with. And good junk yard watchers learn which ones specialize in what. For instance, we have seen grave yards for semi-truck trailers. A wrecked trailer can be a source for some interesting doors and hardware, insulation and flooring plywood, as well as sheet aluminum, which can be used in the same ways as printing plates but with even better results because it is heavier gauge and in larger pieces.

The variety of metal objects in junk yards is so vast that there would be no place to start or stop a list. There is virtually nothing that can't be made out of junk yard stuff with a dose of imagination, labor and proper equipment. Besides structural metal and sheet aluminum you will find ducts, plates, wire-mesh, stands, bases, boxes, cases, culvert pipes, fasteners of all kinds, tanks, covers, access doors, slats, straps, fence parts, gates, cans, hardware and a thousand other types of objects, some of which you will recognize and many more that you could never identify.

■

nary interior design, but it has to be carried all the way without anything soft-cornered or frilly if it is to be pure High Tech.

■

Steel cans and drums (most commonly five- and fifty-five gallon) can be put to a great number of good uses. See the chapters on "Heating, Air-Conditioning and Ventilating" and "Solar Energy Use" for ideas in those areas.

Geodesic dome shingled with discarded aluminum printing plates. The dome contains a bedroom, bathroom and a large closet.

"High Tech" design can be, at once, practical, whimsical, utilitarian, beautiful and inexpensive, if one wishes to be a purist with his project. Junk yard gems, some foam pads, a few pieces of salvaged glass and mirrors and a bunch of metal enamel will produce extraordi-

They can also become floats for boat docks, chairs, table bases, storage tanks, fan housings and planters, to mention just a few possibilities.

Even automobile junk yards can furnish building parts for someone who doesn't mind being different and who has equipment available and knowledge of how to use it.

We mentioned auto glass in the "Glass and Mirrors" chapter, and we have seen some pretty far-out furniture made with remodeled car seats. And there was an episode in a book once about a guy who made a press and jigs which flattened, cut and shaped car tops into triangle "pans" which he fastened together into a geodesic dome.

In a recent issue of *New Shelter* magazine, there was an article about a house (built of a series of silo caps) which was really great. And large, flat-bottomed culvert pipes would make fine subterranean shelters. Special attention would be required to make them leak-proof and well insulated.

There is one important thing to remember about using metal for structural and cladding materials: metal has almost no insulating value and it conducts heat and cold, which means that a lot of insulation is needed between the metal and you. You might have seen walk-in freezers with frost on the outside ends of hinge bolts, which gives an indication of thermal conductivity.

Cables have tremendous strength for their size. This writer once hung a cabin over a creek with a cable running through a pipe at the rear wall, which rested at two points on a natural rock ledge. The cable was ⅜″-diameter galvanized steel, which ran up the hill in back of the cabin and around an old car axle which was drilled into a huge rock outcropping. Two turnbuckles were installed for leveling purposes. Unfortunately, the cabin burned and there are no pictures good enough to include here.

Junk sculpture has sort of gone out of style, but it's still fun and can be functional as well when made into bird baths and feeders, bridges, walks, steps, staircases, lighting fixtures, furniture, etc.

Almost anything metal can be put to some use. The late, great architect Bruce Goff once used aluminum pie plates for shields in front of light bulbs on the interior, front wall of a church in Oklahoma.

Along that same line, we have installed salvaged light sockets, out of some multi-bulb hanging fixtures, in one-pound coffee cans for ceiling recessed lights. Tin discs, with small apertures to concentrate light beams, were installed flat against the ceiling and painted the ceiling color. Care had to be exercised to assure that the wiring would not be cut by the sharp edge of the can-top. Small pieces of flexible conduit are good protection. Another important requirement is to put enough holes in the top to allow heat to escape. And avoid installing insulation over the can.

A friend who used to be in the metal salvage business, David Karchmer, once gave me some brass discs about ⅛″ thick and 3″ in diameter because he thought they were too pretty to be melted down. I built some furniture out of 4 x 4 clear-heart redwood and routed the brass circles flush with the face of the wood. It appeared as if solid brass rods ran through the 4 x 4's, and the contrast of the redwood and brass, which were both finished with clear lacquer, was rather fine, if I do say so myself.

And Dave made window shelves, a bar, light fixtures, hi-fi equipment storage units and all kinds of extremely sophisticated (even elegant) items for his house out of junk he "just had to keep."

You might produce a piece of furniture or work of art that deserves being plated. Look up plating companies in the yellow pages and get bids. Companies that primarily chrome-plate car bumpers have large enough tanks for good-sized pieces.

Metal has value of some amount at the junk yard or recycling center. Copper wire can be burned to remove the insulation, for instance. At the end of your project, gather up all unused metal and beer cans and head off.

CHAPTER FOURTEEN

Glass and Mirrors

Manufacturing techniques have changed in recent years, but since we are concerning ourselves with salvage materials primarily, we will talk about what used to be manufactured.

There is one major exception: reject or mis-manufactured tempered glass. Much more tempered glass is being produced today than was a few years ago because of recent codes and laws requiring, for safety reasons, its use in all door glass and glass panels adjacent to entrances.

Because of the fact that tempered glass cannot be cut, shaped or drilled, when a piece is, for some reason, a reject, it can't be resized and used as a smaller piece. Makers of sliding (patio) doors accumulate large quantities of these pieces which are, usually, 34″ x 76″. They are generally single panes, but sometimes insulating panels (vacuum-sealed double panes, usually ⅝″ thick) are available.

They are rejects for several reasons: discoloration, white spots or streaks, pocks, scratches, etc., which will be visible when your project is finished.

Tempered glass will not break easily by thrown objects, hail, or being bumped against, so it's great for skylights and greenhouse roofs. It *will* break easily (and astoundingly, in a million little bits) if a nail or screw is driven into an edge or an edge rubs against a metal piece, so be careful how you install it!

Other size pieces are sometimes available, but in smaller quantities. So acquire your tempered glass, measure it, inventory it, and design your structure to accommodate your inventory, including plenty of room around each piece to seal and secure it in place.

■

Heavy winds, vandalism, stray flying objects and out-of-control vehicles are constantly breaking storefront plate glass. Some of these pieces are huge, and the breaks are relatively small, leaving sizeable pieces out of which very usable shapes and sizes can be cut. Regular glass (1/16″ single-strength, ⅛″ double-strength, (3/16″ and ¼″) can be cut by anyone with a little practice and a good, sharp glass cutter.

When glass companies are called by store owners or insurance companies to replace broken storefronts, it is their responsibility to remove the broken pieces and haul them away. Used glass (because of scratches, painted signs, pieces of tape, etc.) is not profitable for a glass company to salvage, so they will let this material go either free or close to it to avoid having to pay to have it hauled off.

Warning! Glass, especially large pieces, is hard to transport without breaking and should always be moved resting on edge. Glass companies have trucks with vertically leaning and

*Walls of weathered hardwood scraps with occasional pieces
of off-fall mirror. Designed by author for very special friends.*

padded racks. You might be able to work a reasonable deal to have it delivered. It wouldn't hurt to ask about delivery and cutting, too, since they have big, carpeted cutting tables, and glaziers can handle glass like ball bats without any breakage. It looks impossible and is fun to watch.

◼

Glass is manufactured in standard sizes and cut to customer's requirements. This means that off-falls accumulate. These are usually too small to be used practically, but we have seen some fantastically creative, practical and neat-looking applications of these little pieces in doors, walls, patterned table-tops, mobiles, hand-made light fixtures, etc., etc. Remember that on the north side of a house it's good to have very little glass, thus curtailing heat loss. Small glass areas, however, allow little loss of

energy while adding greatly to the livability of north rooms.

◼

Laminated safety glass, which is two pieces of glass stuck onto either side of a film of plastic, is all over the place, including auto salvage yards. Again, it is hard to cut. Some-times curved pieces are available from older model cars and trucks, cheap because there is little call for it. It might seem pretty far-out and far-fetched, but a good craftsman with routers, etc., or someone with a lot of time, patience, determination and a chisel can do some pretty astounding things with wind-shields and curved glass pieces for protruding windows, furniture bases, light fixtures, etc.

◼

Two pieces of glass installed with a dead-air space between must always be done so that one pane is removable. Dirt will accumulate between the panes no-matter-what!

■

Scraps of mirrors are a different situation! Wonderful things can be done with little pieces of mirrors used in paneled walls (see photo on page 86), installed on walls or ceilings as uniform-shaped tiles or irregular-shaped mosaic, or on tables, cabinet backsplashes, art work, almost anywhere! Mirror can be glued to almost any backup material, but be sure it is solid and won't flex and allow pieces to fall off on somebody or an heirloom. The glue isn't cheap, but it goes a long way and the effects achievable make it more than worth it. Pieced-together mirror is *not* good for viewing one's self, because it cannot be installed as a smooth plane.

Occasionally, mirrors can be used to accomplish passive solar heat gain, on areas such as the north wall of skylight "wells" or on north walls inside clerestory windows which face south. Since sun rays produce no heat until they are absorbed by something, it is good to calculate winter sun angles and install dark-colored masses (like brick) which will warm up during the time of sun reflection off the mirrors. The collected heat will emit from the mass after the sun rays are gone.

■

Stained glass, both new and old, has been used very effectively in "handmade" houses for some time now, and is still very popular. Because of its popularity it can't often be found for free or cheap any more. Keep a lookout for it and grab it if you get a real deal, because it is fun to use and will increase the value of your property. Or you could sell or trade it to push your project ahead, if the stained-glass route is not "you."

Old, salvaged glass blocks are sometimes available and can be used in innumerable ways. If they are not "cleaned" — if the mortar is still on them — be sure you can get it off without breaking the blocks. Glass blocks can be used in walls between kitchen cabinets, for bathroom walls, etc. We have also seen articles on glass blocks for solar-gain walls, and, in one case, the blocks were drilled and filled with anti-freeze water solution to act as heat storage.

■

Glass over ¼″ thick is rare and hard to cut. It is sold for table tops and other specialty items, and available old pieces are usually scratched and chipped. But grab it if you can — you can sand it, etch it with acid, sandblast designs in it, use it for structural supports to make things appear to "float" and even fabricate aquariums out of it.

■

Solid glass doors (without frames) are sometimes available where stores are being demolished or renovated. They are tempered glass, about an inch thick, with holes manufactured into them for hinges, locks, closures, kickplates, etc. Remember: tempered glass cannot be cut, so plan to use them where frames, etc., will cover the holes. These things are heavy as hell, so watch your back.

■

Another type of glass you might be able to salvage is wire-glass. This stuff is used for security enclosures, panels in warehouse doors, etc. It is not very pretty and is difficult to cut and work with. But there are some applications, such as in "High Tech" design.

■

Remember when someone laid up a wall with glass bottles? That caught people's eye so much that some plastic sheet manufacturers began producing trashy-looking "bottle-glass" panels. (We'll look more at plastics in the next chapter.) Building with bottles, it seems to us, was a "cute" idea. The time and mortar required and the difficulty of replacing a broken bottle would discourage us.

■

Insulating glass (still sometimes called "thermopane") is two pieces of glass (now there are some with three pieces of glass) of various thicknesses separated by various amounts of space between. Around the perimeter there is a thin piece of metal and a sealant. Air is drawn out of the inner space to create a vacuum. As with any manufactured product, goof-ups occur. Pieces are made the wrong size or shape, a reject piece of glass slips into production or a leak develops somewhere around the seal. The latter goof is usually discovered after installation when the panels fog up. If you are on the job when the panel is removed by a disgusted installer, you can probably "just take the damn thing." It only takes a little time and a sharp knife to salvage a couple pieces of glass, or one piece if the opposite one is broken.

Insulating glass is fine, but it is not highly insulating. It is *much* better to have single-pane glass on a south wall, for instance, with a closable shutter or curtain of an insulating material to close when there is no sun.

■

Caulking is cheap and a little practice will make you a pro caulker. Use it *everywhere* around glass, to stop water and air infiltration and noise. You'll be glad you did.

Glass can be sanded with a belt or vibrator sander to make it obscure for privacy, light fixtures, art, or just for the hell of it. You can also sand the edges when they will be exposed. But to polish edges, or miter for corner installation, you'll probably need a glass company. Get a bid or two first.

Mitered corners are beautiful and are accomplished with the use of clear silicone adhesive, but it is generally a bad idea to install edges of glass touching each other or touching anything metal (particularly nails) because it will screech and break.

■

More is said about glass in the chapters on windows, doors, lighting, solar energy and furniture.

■

When you go to the glass companies looking for glass, ask about crates that are being discarded. You might put this plywood and lumber to some good use.

■

CHAPTER FIFTEEN

Plastics

Webster defines the noun "plastic" as "any of various non-metallic compounds, synthetically produced..."

As we all know, plastic materials are constantly replacing old, conventional ones in the manufacture of everything from boats to church spires and from clothing to water pipes. The preponderance of newly conceived items are plastic and "we ain't seen nothin' yet."

Some types of plastic materials used in building have been around so long that they have become the "conventional." An example of these is "laminated plastic" (commonly referred to as "formica"), which was introduced around 1927.

Other plastic products were developed, marketed and had their heyday and, for one reason or another, have become less popular. An example is corrugated fiberglass which everyone wanted to use for porch roofs. Installation was not always good, and leaks developed, plus leaves and dirt accumulated on top, which looked terrible from below.

And there were those 4″ x 4″ stick-on, imitation ceramic tiles in bilious colors.

Still other, more recent uses of plastics are revolutionizing entire trades, such as plumbing.

Laminate plastic, while an old product, is one of the finest building materials on the market today. It is fairly simple to work, durable and easy to clean. And now it is available in

sizes from 2′ x 4′ to 5′ x 12′. It can be installed over almost any smooth, clean surface with contact cement per directions on the can. Sharp saws and good files are necessary, and a hand router with two cutters (straight and bevel-edge) will save a lot of time. Cut the plastic a little larger than the area to be covered to allow routing and/or filing. Cabinet and table edges should be covered first, allowing the top to lap over.

Manufacturers of factory-built cabinets make "coved tops," which usually have a rolled front edge and a coved 4″ backsplash. A significant percentage of these coved tops are defectively manufactured and disposed of. The defects are often very minor and easy to work around, and the single piece of plastic (laminated to a ¾″ particle-board sub-top) makes a clean, lasting cabinet top. Cutting ends, especially mitering corners, is tough to do without proper equipment, but getting a cabinet shop to make the cuts for you shouldn't cost much. Ask the price first and ask, too, how much it would cost to have pieces routed off where broken on edges, etc.

If you don't live near a cove-top manufacturer, you will probably find reject tops at salvage stores and surplus retailers. The cost will be much less than the price of the materials which went into them.

Reject sheets of laminate plastic are not likely to be available unless you can get to one of the manufacturing plants. The *Thomas*

Register, in the library reference department, lists the various manufacturers and their plant locations. Almost all laminate plastic is manufactured the same way, to meet government standards, so one brand is as good as another. The sheets used for coved tops are thinner, and a little less durable, than standard ¹/₁₆″ thickness, and there are new products out now which are the same color all the way through.

■

Acrylic plastic (plexiglas) is formed into many shapes of many different colors. Clear sheets of ⅛″, ³/₁₆″ and ¼″ are most common, but tubes, bars, rods, "bubbles," and many other shapes are manufactured in numerous sizes.

Most acrylics are easy to cut, sand, polish, drill and glue, but some, such as bullet-proof sheets, are not.

One disadvantage with most acrylics is the way they get scratched up. Another problem for the scrounger is that, because it is expensive, uses are found for off-fall pieces down to pretty small sizes.

In the case of tubes, rods, etc., however, there is not a great demand for short lengths.

Most medium-sized and larger cities have plastic products companies which sell acrylics and fabricate specialty items. These places are listed in the yellow pages under "plastics."

Pieces of clear tubes used as furniture legs create the illusion of the piece "floating." We have recently used 4″ diameter tubes in place of the heavy wood legs under a parlor-grand piano. Larger diameter tubes (12″ and 16″ diameter) are great for table bases and terrariums and aquariums. Any size or color tubes will make good light fixtures, display pedestals for objects of art, umbrella stands, planters and, stacked like cordwood, storage compartments and wine racks. Street address numbers, trivets, trinket shelves, drawer dividers and cabinet doors can become extra special when made of plexiglas.

And this stuff has a quality not found in other materials, which can be taken advantage of with very striking results: it is a great conductor of light from edge-to-edge.

■

Fiberglass is still manufactured in the aforementioned corrugated sheets (used a lot for skylights in metal buildings), flat sheets of various thicknesses, chairs, shower enclosures, bathtubs and whirlpool tubs, special doors and uncountable products which are non-building-related.

Manufacturing techniques are constantly improving, and this results in better quality fiberglass products. Several years ago, shower stalls tended to be porous and flimsy. For this reason, you are not likely to pick up a shower stall that is worth taking home.

Even the big-name bathtub and whirlpool manufacturers (American Standard, Jacuzzi and Kohler) are using fiberglass now because some of their products are so large that they could not be man-handled if manufactured of cast iron.

Plants which fabricate fiberglass boats, tub-enclosure doors, pick-up camper tops, chemical toilet buildings, overhead doors, etc., all produce some rejects. Find a plant near you and inquire what it takes to procure some.

■

As mentioned back in "Insulation," when acoustical tile ceilings are being removed, so, usually, are fluorescent light lenses.

These plastic gems are 2′ x 2′ and 2′ x 4′ pieces of one of several different white plastics. Except for the really cheap ones (which will have turned yellow), they can be cleaned and reused for lighting or used as panels in cabinet doors, shoji screens, table tops, room dividers, bathroom windows, etc.

Most of these lenses will be about ⅛″ thick, smooth on one face and patterned or sanded on the other. Some, however, are "egg-crate" —

about ½" thick with thin, white plastic strips running both ways and forming a panel of holes approximately ½" square. The possibilities for their use are so varied and fun that we are anxious for you to send a picture of what you do with them.

■

Thin plastic sheets and films (visqueen) have been discussed in earlier chapters, but are worthy of further mention.

Plastic lenses in wood grid for illuminated kitchen ceiling. Fluorescent "shop-lights" are often on sale really cheap and are perfect for this use.

Light lenses are often found in fairly large quantities, allowing the finder to use them liberally. One wonderful use is to create a fully lighted ceiling in a kitchen or bathroom. If you get the "T" bar system with the panels and lenses, you can repaint and use them for your illuminated ceiling or, as the photo above illustrates, you can hang a wood divider system in place. We have found that about seven feet is a good height for an illuminated ceiling, and this works perfectly in remodeling a room with eight-foot or higher ceilings.

Two-, four-, and six-mil polyethylene plastic sheets are available in black and clear and in pieces up to twenty by one-hundred feet. This stuff is inexpensive and indispensible for use as vapor barrier (see chapter on "Insulation, Vapor Barriers and Sealants"), covering over earth in crawl-space below floors, water-stop below concrete slab floors, protective covering for materials, temporary window and door-opening, open-wall and unfinished roof-area protection. And until you can afford storm windows, it works quite effectively tacked with strips over windows, but it isn't very pretty and it distorts the view.

Heavier gauge films are available, but at greatly increased prices. Some of these products significantly reduce heat loss/gain and are workable (using a hair dryer, etc., for elimination of wrinkles) so that there can be minimal sight distortion.

The use of plastic film for greenhouse wall and roof covering and solar collection glazing is feasible, but it cannot be expected to last more than a couple of seasons, at best.

■

Plastic pipe is manufactured for use for sewers, lateral fields, storm water perimeter draining, gas, cold and hot water. But, like anything else, plastic pipe does not have to be used for the purpose for which it was designed. Do watch, however, not to use old gas or sewer pipes for water piping. It probably wouldn't kill you, but the taste would be *bad!*

Another good idea you might come up with, this writer already beat you to — with unfortunate results. Several years ago in Kingston, Jamaica, a friend and I cut some 4″ PVC pipe in two, lengthwise, developed a brilliant system to install it, and had a roof laid like Spanish tile, except in full-length pieces. It was beautiful — for a few months — until the sub-tropic sun disintegrated it.

Using one's imagination and own hands to create something great from very little is satisfying beyond belief. But there is always danger in straying off the beaten path, so don't expect every single brainchild to be genius.

■

Foams, in sheet and block forms, vary widely in density, degree of firmness, shape and size.

There is foam rubber, polyurethane, polystyrene and various other foam. Reject foam is available at manufacturing plants. Damaged foam and small pieces are available, at discount prices, at furniture factories, upholstery supply outlets, mattress factories (there are probably some small ones in your area), and upholstery shops. And, for some reason, salvage and surplus stores usually have big piles of foam, but it's not necessarily very cheap.

Much fragile and sensitive merchandise is packed with pieces of very usable foam. Video equipment, for instance, is often shipped in boxes containing two sheets of good quality foam about 1½″ x 16″ x 20″.

Pre-used foam can be found in all kinds of places and is usually in a lot better condition than you might think at first sight of it.

Unlike ordinary mattresses, foam ones are usually in good shape under the tattered cover, and can be found in used furniture stores, auctions and garage sales. Watch sizes, if you need a specific one. Foam pads out of recreational vehicles are usually smaller than standard mattresses.

Much discarded furniture will yield usable pieces of foam as will seats out of wrecked cars. Foam carpet pads are extremely good for some applications and can be found at demolition and remodeling sites.

Four-inch thick foam rubber on a solid platform is about the most comfortable and healthy bed made. We know of a bedroom which has a six-inch thick foam floor made of alternating layers of 1½″-thick poly-foam and ½″ foam carpet pads. A soft, pliable carpet covers the foam, wall-to-wall, attached to strips of wood nailed to the walls. The owners spread king-sized sheets and blankets for their bed.

Padded table bases and furniture platforms are sort of "in" at the time of writing, and padded head boards for beds make good sense for backrests while sitting up in bed and as added insulation between you and the outside wall.

And there is the possibility of using foam for built-in seating — an example of which is shown in the photo on page 93. This is a ten foot long sofa which is suspended from the wall by steel cantilever brackets. Layers of foam of various density were glued to pieces of ¾″ plywood to form the seat and bolster. Synthetic plush upholstery material was stretched tightly over the foam and attached to the bottom of the plywood, contouring the foam. Wooden wedges were installed between the plywood seat-bottom and steel brackets to make the seat 1½″ higher in front, as required for sitting comfort. The bolster is likewise sloped back at the top.

There are, of course, many more kinds of plastic materials than we could think of, and any you find can probably be utilized in your project with sufficient thought. We would always appreciate pictures of your work.

A very new plastic building material is called "Corian." We mention this stuff because it is manufactured in sheets and cut into smaller pieces of various shapes. This cutting and shaping is performed in fabricating shops and architectural millwork shops.

When "making little ones out of big ones" occurs, off-falls usually develop. Small pieces of Corian can be recessed into kitchen cabinet tops for setting hot pans on and using as cutting boards.

Because Corian can be milled, cut, sanded, glued and drilled, it can become all sorts of items from door pulls to art work.

*Foam rubber on plywood on steel brackets attached to wall posts form this
ten-foot-long sofa-bolster cushion which is hung onto wall on window screen hangers.*

■

CHAPTER SIXTEEN

Doors and Pieces of Doors

Regular, hinged, walk-through doors are generally one of the following types of construction: hollow-core, solid-core, flat-panel-stile-and-rail, raised-panel-stile-and-rail, louvered, glass-paneled or combinations of some of these. There are also sliding (patio), dutch, bi-folds, by-passing, steel, aluminum (storefront), solid glass (see chapter on glass), overhead, barn, fire-retardant, antique and weird doors.

We'll try to cover types, availability, advantages/disadvantages and uses as much as possible, but only the limits of your imagination will determine where this chapter ends.

First-off: hollow-core and solid-core doors manufactured for residential and commercial building are sold by lumber yards, architectural millwork companies and millwork wholesalers, all of whom damage some of their products with forklifts, improper handling, dropping, climbing onto, stacking, etc. Damaged goods take up valuable space to warehouse, so they must either be repaired (very often too expensive to be feasible) or gotten rid of.

Most residential interior doors are 1⅜″ thick, 6′ 8″ tall and 2′ 0″ (closet and bathroom), 2′ 6″ (bedroom), 2′ 8″ and 3′ 0″ wide. There are many other standard sizes, from 1′ 0″ up, but the former are the ones most commonly available. These are hollow-core, which means that they are two very thin pieces of plywood glued onto either side of an incredibly small frame with as near nothing inside as anyone can figure out

how to use to keep them from collapsing of their own accord. They have very little thermal or sound-insulating value and are extremely easy to puncture. The thin face veneer is usually mahogany, ash, birch or a pre-finished (sometimes photographed paper) film. Unlike the solid-core doors discussed below, these doors have little application as table tops, etc., because there is virtually nothing in them which will hold screws.

Damaged hollow-core doors are, however, not completely useless. Holes punched in them are practically non-repairable, but sometimes something light can be glued over the hole. Don't try mirrors unless you can fasten them clear out on the outside edge — the frame stiles are only slightly over an inch in width. When only one face of hollow-core doors is damaged, they can be used for paneling, for doors into little-used areas where a hole won't be seen much *or* they can be taken apart. If a table saw is set up to cut the thickness of the plywood face and a cut is made all around the edges of the plywood skins, they can be pulled off the frames, rendering some nice pieces of plywood about ⅛″ thick for use as paneling, to veneer furniture areas, etc.

Twenty years ago this writer came into possession of thirty pieces of hollow-core doors which had been cut (to allow installation of glass panes) from doors used at classrooms in the Sunday-school wing of a large church. They produced sixty pieces of ⅛″ thick birch plywood 26″ x 36″. The photograph below shows how they were used for interior paneling, with ¼″ x 1⅛″ yellow-pine battens and, also, how new 2′ 8″ x 6′ 8″ x 1⅜″ hollow-core doors were stopped-in between the 4 x 4 posts on 3′ 0″ centers which form the frame of the post-and-beam structure.

density wood-chip board (which is not desirable because it won't hold screws, nails or glue well). Doors manufactured for commercial buildings are "architectural-grade" with higher density cores or lumber cores. You are not too likely to find reject fire doors (they have UL

Hollow-core doors, and pieces cut from doors for
glass openings, used for walls in author's house.

Some exterior residential doors are also, regrettably, hollow-core, but they are 1¾″ thick with a little more back-up. Generally, the above applies to them as well.

Solid-core doors are a whole 'nuther ball-game! They are more often 2′ 8″ and 3′ 0″ wide (sometimes even wider), 1¾″ thick (sometimes 2¼″, but not often), 6′ 8″ or 7′ 0″ tall (occasionally taller). The residential solid-core doors are normally 6′ 8″ high, of the same wood species as hollow-core and cored with rather low-

labels on one edge that designate them as "B" or "C" label and are cored with gypsum, like sheetrock), sound-retarding doors which have a not-so-good soft-core (they are good for kid's playroom doors), or x-ray area doors, which you will immediately recognize when you attempt to lift one because they have a layer or two of lead in them. Commercial doors are manufactured to architect's requirements, so you might find faces of anything from knotty-pine to rosewood. White ash, red oak, white oak and birch are probably the most common at the time of writing.

Again, residential doors are handled, therefore damaged and disposed of, by lumber yards and their wholesalers. Commercial doors are sold by architectural millwork companies and some building specialty companies. These people can be found in the yellow pages, but often, because they do not sell to the general public, they are in there in small type without ads. It is simple enough to call any company to inquire if they have accumulated any damaged or reject materials, and they are often more than pleased to be able to unload some. Please remember, however, that it costs any employer a lot of money for someone to show you through his stock, so don't go looking until you are ready to begin to accumulate stuff.

Hollow and solid-core doors are manufactured in plants all across the country and they all have rejects. Your library reference department has books listing manufacturers of everything. There may be a door plant near you.

Holes for glass panels are more often cut in solid-core doors than hollow-core because it is easier to install the stops that hold the glass in. These holes are sometimes cut at the plant per "shop-drawings" and other times at the millwork companies, so they are available, sometimes, in very large quantities, from either source. Sizes of these cut-outs vary enormously, from 4" x 8" or smaller to 26" x 64" and larger.

Except for the very small ones, solid-core door cut-outs are nothing less than wonderful for the person building himself a house out of junk and stuff. The corners must be trimmed a little with a hand-plane, jointer, saw or belt sander before they can be recut on the table-saw or banded, but that's no big thing.

Banded with thin strips of wood or plastic-laminate (formica), they can be table tops, furniture ends, shelves, divider walls, benches, platforms — the list is endless.

They also work beautifully for cabinet ends, dividers and shelves. With 1¾" face frames on cabinets and "full-overlay" doors and drawer fronts, for instance, drawer slides and adjust-able shelf hardware can be mounted directly to the cut-outs with no filler blocks. Left-over pieces can be used for the base "chairs" (front-to-back pieces below cabinets) which should be cut short an inch or so at the back to allow the cabinet to fit tight against the wall when there are irregularities where wall meets floor. It also helps to cut them "arched" at the bottom so floor irregularities won't cause the cabinet to rock. We have even seen them used for doors and drawer fronts, either full-overlay or partially inset, but door hardware might be too hard to find or too expensive for this to be practical.

Formica over face and exposed edges, if you like, but seal the opposite face to prevent warping due to the unfinished side absorbing moisture and expanding. If you use drafting board linoleum, do the same thing. And the same goes for sheet vinyl or vinyl stair tread material or carpet.

We have some architect friends — John McWilliams and Dennis Spencer — who have built themselves, in stages, all of their drawing boards, storage units, bookcases, etc., out of damaged solid-core doors and cut outs.

∎

Stile-and-rail doors are the old-fashioned, exposed-frame doors, still manufactured, with various panels, glass pieces, louvers, etc., installed between the stiles (vertical members), bottom rails, lock rails and top rails (horizontal members).

S & R doors manufactured today for residential interiors are usually fir. Residential exterior and commercial S & R doors, such as store-front doors (one large glass with no lock rail) are usually white pine.

Panels, whether ¼" plywood or hardboard (masonite) or raised-panels, are manufactured into the doors with no removable stops. This means that to replace one you would need to cut out the milled edge all around the panel and replace the removed portion with some kind of little stops — which is a real pain! And replacement raised-panels are not available and very difficult to duplicate.

Louvers are even worse! It might look simple enough to make a new little slat or two and slip them in, but it *ain't*.

Glass panels are usually installed with a removable stop on the inside of the door, so that's fairly simple.

Generally speaking, damaged S & R doors are not so good to salvage. The 1¾″ thick ones will produce a few pieces of nice wood, and you can always cover broken panels with salvaged pieces of plywood, hollow-core door skins or something you might have been able to scrounge. Insulation can even be put between overlaid panels for a pretty good entry door, but you'll spend a good bit of time making them presentable.

Damaged louvered doors can often be cut down in size, to eliminate the damage, and used for doors on cabinets, linen closets, storage units, etc.

Broken stiles and rails are most often broken "with the grain," and with some care, white glue and clamps you can repair them so they can be filled with "rock-hard-putty" or wood filler, sanded and painted. Be sure to use blocks of wood any time you clamp something, to keep the clamps from denting the wood. Forget trying to fix a break which goes across the grain of any piece of wood.

Our advice concerning buying damaged S & R or HC doors is this: if you can pick them up for a buck or two, or for hauling them away, you can probably find a way to use them.

■

In most tract houses and a lot of custom houses and light-commercial buildings, pre-hung door units are being installed. These units include the door (or double-doors, bi-fold set, etc.), drilled for lockset and hinged to a frame, with integral stop, and with the casing (trim) applied to one side and loose, to facilitate installation, on the other side.

Producers of these units would not be likely to sell rejects, because damaged pieces can be easily exchanged. We have, however, known of derailed freight trains and overturned trucks carrying many pre-hung door units. When this type of thing happens, the railway company or truck line sells the whole load, damaged or intact, at auction or to salvage stores or surplus stores.

Generally speaking, this is pretty cheap, common stuff. If you get a chance at it, be sure you can use it and want it and that you don't pay too much for it. Know the going retail prices.

■

Nowadays pocket doors are usually hollow core, hung from a light, steel track on rinky-dink nylon rollers, and are installed in a "pocket frame assembly." This is a light metal frame with furring strips of wood on each side; it takes the place of a stud wall where the door needs to roll into a "pocket." These are trouble-prone at best (and hard as hell to fix, if anything goes wrong), so we can't imagine anyone trying to use a damaged or pre-used unit.

Most bi-fold and by-passing hardware is also pretty flimsy, but at least you can get at it to adjust or repair fairly easily.

■

Old hollow-metal doors are often available at the demolition site of commercial buildings. Unless they are insulated (unlikely) don't try to use them as house entry doors. They, with their HM frames, are good to use as entry doors to unheated areas or buildings, or they could be used inside, but they aren't very pretty.

HM doors will often make decent work bench tops. Some have excellent fire ratings. You might use them for a protective enclosure around a wood furnace or something of that nature.

■

Everyone is looking for neat, old doors with stained, etched and beveled glass or carved panels or overlaid, carved or milled trim, etc. When everyone is looking for something, the price goes up. This is not to say that there are no real deals anymore, but they are harder and harder to find.

Occasionally, still, on a drive along country back roads, you might find an old, abandoned house or country store building with some goodies still intact enough to be salvageable. We have found that it is often difficult to find the owner but it's always worth a try.

Antique doors were very often hand-made, and, therefore, in some very unusual sizes. It is always better to acquire your materials, measure and inventory them so you can plan your structure to accommodate them. Old doors have also been fit to out-of-square frames and trimmed repeatedly as the old building settled, and they have been mortised, drilled and cut-on for non-standard hinges, locks, dead-bolts, etc. And the older the door, generally, the lower the doorknob was installed. So be prepared to spend some time patching, filling, squaring, sizing, refinishing and finding hardware and cover-plates for these old treasures.

Remember, when you find something old and valuable that you feel you don't need to use, you can sell it or trade it for something you do need.

■

Sliding patio door units are manufactured almost exclusively out of aluminum, but there are some wooden ones. The aluminum units range in quality from pretty sound and solid to almost trash. The newer ones all have tempered glass, because there is now a law which requires it. This glass is discussed in the chapter on glass. Older units may have ordinary glass in single pane or insulating glass. But glass is not what determines quality. Thickness of the extruded aluminum, size of track, connections, seals and rollers are what determine quality. Look them over closely and try to examine them before they are removed from

their present openings, if you find them being removed.

Screen panels don't function too well, at best; because they are so light and flimsy, they are hard to keep on their tracks. Locks are, also, usually inadequate.

Recently we found two 6-foot units, which had been manufactured by P.P.G. around 1959, where a medical building complex was being removed for the sake of "progress." These units, which we bought for $50 each, are some of the heaviest and best ones we have ever seen. Some real bargains are still out there!

Wooden units are great, but you are not likely to find them being removed yet and it is doubtful you could find any damaged ones.

■

Overhead garage doors are, in our opinion, one of the very few bargains in new construction today. Consequently, salvaging, repairing and hanging one yourself may not be worth the hassle. This is especially true of a unit which is fifteen years old or over, or one that is not running on its tracks smoothly or opening and closing fully.

If you do find a "steal" on one, the best advice we could give you is to find an installer for one of the major overhead door distributors and talk him into reinstalling it for you, on his own time, to save paying company overhead and profit.

For many years now, overhead doors have been made in sections — four, five or six horizontal sections hinged together — which roll up and back on steel tracks. Some older models were single-piece doors which pivoted out at the bottom and rolled up and back on the tracks. Don't even think about trying to salvage a single-piece door for reuse as a garage door. But you might figure out a way to use that big hollow-core door (about 7' x 6' 8") for something else.

Most sectional door units are wood stile-and-rail construction with hardboard (masonite) panels and one section with glass panels. What we said about S & R doors above pretty well holds true for these in terms of salvaging parts for other uses. Some are hollow-core, for which you could probably come up with some interesting alternative use of the sections.

Then there are some made of fiberglass, aluminum and light steel. It seems to us that sections from these could be used effectively as either stationary or moveable, horizontal or vertical louvers at overhangs, in front of large glass areas, etc., to control sun rays in the various seasons of the year. Manufacturers (such as the Overhead Door Company, Franz, etc.) have reject panels that may be available if you live near one of their factories.

Dutch doors are worthy of mention here because they are quickly discarded as unusable when removed at construction demolition sites. Because they are two doors (one above

the other), they have a little more reusable hardware and often a formica or wood shelf. Reuse of a dutch door in a house can sometimes be a very handy feature in a home where there are toddlers. The top leaf can be left open so the little ones can be watched and heard, with the bottom leaf closed to protect them from the basement stairwell, etc. As a matter of fact, two severely damaged doors can be cut in two to form a dutch door.

If you just salvage the materials from a dutch door, the above comments on HC, SC and S & R doors apply.

There are many, many types and styles of doors we have not covered here because they are not plentiful, space wouldn't permit and we don't know them all. But doors, on the whole, are one of the most fun types of materials to salvage.

Architectural office furniture fabricated from
damaged solid-core doors and discarded pieces.

CHAPTER SEVENTEEN

Hardware

It probably makes sense to follow the chapter on doors with hardware, since finish hardware is mostly attached to doors. There are, however, other finish hardware products for, most commonly, cabinets and windows.

We refer to "finish" hardware to differentiate it from "rough" hardware (such as anchoring devices, post bases, etc.), which is discussed elsewhere.

Except at going-out-of-business auctions at lumber yards, hardware stores, etc., you are not likely to find new hardware other than from normal retail dealers. Occasionally surplus stores get some in, but it is important that you familiarize yourself with going-rate prices before buying at the surplus store or bidding at an auction. We have seen lots of stuff (particularly tools and hardware) go for more at auctions than the price, off-the-shelf, down the street.

Also, especially with door locksets, cheap hardware is just that. Cheap key-in-knob locksets don't last long and the plating wears off the knobs quickly.

There are two different configurations of loose-pin door hinges (butt hinges), and it's important you know what you're getting! They're either square-corner or round-corner. Manufacturers of pre-hung door units, mentioned above, and builders who hang any quantity of doors use electric routers and mortising jigs to prepare door edges to receive hinges. The router bit does not, of course, produce a square interior corner, so to avoid having to cut the corners out square by hand, round-cornered hinges are used.

If you are going to hang five or ten doors, you are not going to buy a door-routing kit. You will use a hammer and sharp chisel. Did you ever try to cut a round interior corner with a chisel? And round-cornered hinges look really tacky in squared-out mortises.

"Locksets" is the word used for door opening devices (knobs or levers) whether or not locks are involved. They, like hinges, are commonly installed with electric tools and an expensive little device called a "boring jig." You don't need this stuff for a few doors, but a variable-speed electric drill and a set of bits, including flat "spur" bits of the sizes required for your various locksets (usually 1″ and 1¾″ for new locksets) will be tremendously helpful throughout your entire project. There is nothing wrong with the good old hand brace and you can buy an adjustable bit for it.

Entry locksets are varied in type and style, from key-in-knob to fancy handle sets with thumb latches, lever handles, etc. Dead bolts are installed remotely from the lockset and usually have a key slot on the exterior and a turn-latch inside.

Double security sets have a key-in-knob with a dead bolt above in the same escutcheon (cover) plate.

Dummy knobs (for closets, etc.), passage sets (common interior doors), privacy sets (when you want to keep the little ones out of your bedroom, etc.), and privacy bath-sets are commonly available.

New locksets, hinges, etc., come in several different colors. Bright brass is generally the least expensive. Privacy bath sets have a white-metal knob on the bathroom side.

■

So much for new stuff. Where houses or buildings are being demolished, there is gobs of hardware. If you can get to it in time, before demolition begins, you can get entire units of hardware, screws and all. There is nothing we can think of more frustrating, when building with salvaged material, than finding yourself missing half of a hinge, a hinge pin or a piece of a lockset. It is very difficult, and sometimes impossible, to find exactly what you need for a replacement part.

Try, too, to get the strike plates (the pieces on the jamb which the latch or bolt fits into in the closed or locked position), because it is a hassle to have to find one somewhere that will work.

Contract hardware companies do have trays of spare parts, if you can get them to look for you. It's worth a try. If keys are not available (they usually aren't), locksmiths or contract hardware companies can help you (look in the yellow pages).

There is some degree of standardization in hardware which has been manufactured over the last fifty years or so, and you are most likely to find hardware made within that time span.

Commercial buildings coming down yield heavier and better quality hardware, usually. Hardware taken off of metal doors can be used on your wood doors by replacing the bolts with wood screws of the proper size. Commercial buildings also have such good things as slide bolts, panic hardware, security locks, kick-plates, etc.

This writer, having an aversion to door knobs, used heavy-duty magnetic catches (discussed later in cabinet hardware) and wooden pulls, which are part of the design motif, on all interior doors. Small slide bolts were used for privacy in the bathroom and master bedroom. He also installed double doors at his main entrance which are 2' 4" wide. By doing so, a big opening is available for moving the piano, etc., in and out. And it looks good, but there is more uninsulated area for heat escapement.

■

As mentioned repeatedly throughout the book, antique stuff has become very popular, thus expensive. Hardware is no exception, as you know if you have visited flea markets lately.

But there are fantastic things hanging around on old, abandoned houses and buildings out there, and they are sure worth searching for!

Butt hinges were often cast with designs on the exposed faces, and fancy knobs were formed on the loose-pin ends, with matching ones attached to the bottom of the hinges.

Entry locksets were sometimes incredibly beautiful in either simplicity of design or ornamentation.

Metals used were heavy stuff — brass, bronze, copper, pewter, cast iron, etc. Soaking in a cleansing solution and a little hand-polishing and scrubbing with an old toothbrush will do wonders. Of course these things will need attention to keep in this condition, but it may well be worth it.

Knobs can be of almost any metal, or you might find onyx, rosewood, ceramic or crystal glass!

Many old buildings and houses (built before air-conditioning and low ceiling heights) had transom panels above the doors, both interior and exterior. The transom panels themselves, if they are still reasonably intact, are wonderful for windows, cabinet doors, furniture ends, etc., and they often have fine hardware like slide-rods and knobs. These slide-rods can be used for hinged-out windows, covered-bookcase doors, insulated window shutters, etc.

Now, locks for antique hardware can be a real problem. Skeleton-key locks won't catch it in today's world, unfortunately. And the old, non-standard-sized lock cylinders are often unfixable or missing. A bit of study of each individual problem will almost always result in a solution. Remember that for every design problem there is an infinite number of solutions, so what you're looking for is the best you can find in terms of cost and appearance. Again, you can take your hardware parts to a lock-and-key shop or a contract hardware supplier for help.

When the door you come up with for an opening and the hardware you find are not compatible, the answer is often an escutcheon plate which is large enough to cover all of the existing holes or filled-in places and accommodate the hardware. Sometimes it is fun and practical to juxtapose very sleek and very smooth finished materials on an antique door, perhaps, and behind ornate hardware. A plain piece of black plastic, sheet copper, etc., could be cut and drilled to form a perfect escutcheon behind Nineteenth-Century cast bronze hardware.

It is probably worth mentioning that you might happen to come across pocket doors in old houses. Sometimes they are single, but more often double doors between drawing and dining rooms, etc. They seldom work any more, because the house settled too much or they jumped their tracks long, long ago, and it was virtually impossible to get into the track/roller assemblies to make adjustments or repairs. If you find some of these, maybe you should figure out a way to reuse them with trim and/ or a panel on one side removable, which will allow you to make adjustments when you need to, down-the-road. These doors also have interesting recessed pulls and, sometimes,

locks which can be reused with the doors or used for some different application.

Kick plates (usually so cruddy-dirty, scratched and stained that you tend to overlook them) are sometimes available. Turned over and cleaned, they will provide escutcheon plates, furniture panels or some other good things.

Old, double-hung windows often had unique latches which can be used on new (or newer) windows. They can also be used on cabinet doors and furniture. Even the pulls screwed to the bottom rails of some of these old windows are wonderful finds for use in cabinet and furniture building or for walk-through doors if heavy-duty cabinet catches are used in lieu of regular locksets.

Keep your eyes peeled for other hardware items from antique structures: fireplace dampers with protruding brass regulator handles, coat hooks, etc. While not exactly hardware, you should also watch for grilles and grates in floors, walls and ceilings. Some have adjustable-flow mechanisms which will work when cleaned-up and oiled. We find that a passive solar house can work better with openings and closable grilles spotted around in strategic locations to allow heated air to get where it is needed and displaced, cooler air to get back to the solar-gain source. Fans and duct work are discussed in other chapters.

Remember, if you can pick up some of this old stuff cheap or free and you don't have a need for it, take it to a flea market in the city or an antique auction, or sell it through an ad in the paper.

■

Old houses don't provide much cabinet hardware, because they don't have many cabinets. Old general stores, hardware stores, dentist offices, etc., do.

When antique cabinets are available and will come out all in one piece, or in large sections, they should be salvaged intact. When that is impossible (because they were built as a part of the structure by the finish carpenter), the doors, hinges, shelves, bins, cubby-hole units, drawers and such should be salvaged. If none of that works, the cabinet pulls, knobs, latches, elbow catches, locks, roller assemblies, catches (if they still work), slide-bolts, hinges, hold-open devices and adjustable shelf pins can be salvaged and recycled beautifully.

Some "modern" cabinet hardware from structures built after WWII may not be worth fooling with unless you have a particular use for it. Some of these newer pieces are, however, chrome-plated-brass, solid brass or aluminum, and have some value.

Magnetic catches, unlike some of the other cabinet door catches, are usually worth salvaging. They are made in light-duty size with two magnetic plates, and heavy duty with three or four plates. If they don't work when you find them, it is almost certainly because the magnetic plate edges and/or the door plates have been coated with varnish or the catch has been pushed back where it no longer touches the door plate. Just scrape the metal off with a knife and the catch is good as new.

■

CHAPTER EIGHTEEN

Windows

There are several kinds of windows and we'll try to discuss all of those which you are most likely to find when scavenging.

New double-hung wood windows are manufactured in plants ranging in size from one-man operations to giant factories, and they all produce some rejects.

In the trade, a "window" is a pair of sash (frames) that make up a window. A "window unit" is the complete thing, with frame, some kind of weatherstripping or weight-balancing device, and the exterior trim.

Small shops often make unusual sizes of replacement windows and window units to fit custom openings. Any time non-standard sizes of anything are made there are going to be size errors made, so odd windows (sash) and window units are forever being disposed of. We have never heard of larger plants selling rejects, but they probably do.

Besides double-hung windows, made of wood, there are also casements (hinged on the side), awning (hinged top or bottom — usually top), fixed sash (non-opening), picture (large non-opening), sliding, bow and bay windows, not to mention storm windows. There are also windows of all these types, made of wood with vinyl plastic cladding of various colors. They are handled by lumber yards and wholesale millwork suppliers who damage some of them and receive some in with "hidden" damage from freight companies. As with any other pro-

ducts, damaged windows must be repaired or disposed of quickly so that they do not occupy valuable warehouse space.

Some wood windows are manufactured with the glass not removable without cutting the milled bead away from one side. Replacement requires stopping the new glass in with very small wood stops (as mentioned before in the section on stile-and-rail doors). This is no small task.

■

When you install your wood windows, take care to get them very plumb and square to minimize air infiltration and binding. And, as with any window, be sure insulation is installed all around the perimeter and that joints at trim-against-wall, etc., are well caulked.

■

Used wood windows are often salvaged from both old and newer houses and from some commercial buildings which are coming down to make way for shopping centers and eight-lane expressways. The old timers were installed with sash cords and iron weights. The cords used to be attached to the sides of the sash in routed areas, run up the inside of the frame to a hole near the top of the jamb and attached to the top of an iron weight (or weights) which ran up and down in a boxed-in area between the window frame and the cripple stud. We use the past tense because the cords have almost all broken and been cut away, and the weights are resting where they fell on the bottom plate of the wall.

Many of the old windows had sash-lugs. These are shaped extensions at the bottom of the stiles of the upper sash to keep that sash from dropping flat against the window sill when lowered. They give the window some antique snob appeal.

These old sash-weight windows have two grave disadvantages, even if you find some that are not rotted out: For one thing, they have no weatherstripping of any kind, and secondly, the "chases" for the weights to travel must be left open, which means they cannot be properly insulated.

Newer double-hinges have aluminum or plastic friction or spring-balances on the jambs which are quite air-tight.

■

Sliding windows are *not* always air-tight. The other types of wood windows, if well built, probably are. Close inspection will probably tell you if they are good enough for your use and what it will take to repair them.

■

Glass and hardware are discussed in those chapters, but one comment about installed glass should be made. It is virtually impossible to tell the thickness of glass, or even if it is single-pane or thermopane, when it is in a sash. The only way to tell is to measure from the glass to the outside of the sash on each side and subtract that sum from the thickness of the sash.

■

People want stained, etched and beveled glass windows of any design, style or colors.

The most valuable shape is probably the round, "rose" window because of the difficulty of producing it. The next are probably the dome and arch-head windows. But any fancy-glazed antique window is valuable even if it is etched with the words "In memory of our dearly loved father Alfred Newberry." If a bargain comes your way, grab it! Then figure out how to use it, trade it or turn it into cash sorely needed for your project. Then don't forget old Alfred in your bedtime prayer.

Old churches have interesting windows of various materials. Even if the stained glass has been removed, you might figure out how to use a round, dome-head, arch-head or other irregular frame (or a series of them), if they are not rotted out and would hold fixed glass, wood panels, smaller windows or combinations fastened inside with wood stops.

■

Be careful about buying new stained glass. If you like it, fine. But see if it is, indeed, leaded (not made with zinc dividers) and that it is solid and professionally executed. Some new stained and etched glass is excellent and beautiful art but, it does not have the value of antique work even if it is better.

■

Aluminum windows have been manufactured in such a vast range of quality that we can't say much but to be careful when purchasing any. Some are made of such light aluminum extrusions that you can almost bend them with your fingers and the joints will sometimes just fall apart. Mobile homes almost invariably have this quality. Then there are some which are better, some of which are pretty good while some are heavy-duty dandies. Since the first energy scare, several manufacturers have begun producing aluminum units with built-in thermal barriers of various kinds.

■

Self-storing aluminum combination storm and screen windows are, in our opinion, pretty fine things for a lot of applications and can sometimes be found slightly damaged or saved from the wrecking ball.

■

Aluminum storm panels and screens are being made in little shops all across the country, in any size desired, for what we see as very reasonable prices. Because they are fairly reasonable, and they look good because they have such a small profile, buying new might not be a bad expenditure.

This is particularly true when odd shapes and sizes of openings develop as a result of using and combining scrounged materials. If you do find some used ones, however, they can likely be used for some purpose.

■

Many aluminum jalousy windows were manufactured and are available for recycling. These are the windows with the horizontal, glass shutters that crank open and closed. They let air whistle through so badly that they should not even be used with storm windows in a house. If you are installing a screened-in porch, or building an unheated structure or room, however, they are ideal. You can mount screens on the inside with holes through the screen for crank handles.

■

You might find steel windows, still not too rusted to use, in old factories, schools, etc., slated for demolition. These are usually either fixed-glass, hinged-vent, or a combination. Steel is a very fine conductor of heat and cold, so steel windows are not great for energy conservation. However, for greenhouse walls, storage buildings, unheated garages and the outside walls of air-lock enclosures, they might be just right.

■

Remember that windows are openings in walls which allow the flow-thru of air and/or light — so almost anything can become a window from a ship port-hole to a car trunk-lid, or from a scrap of plexiglas to a World War II bomber machine gun bubble. Windows can be the most fun parts of your house.

■

CHAPTER NINETEEN

Wall and Ceiling Materials and Finishes

Almost every preceding chapter deals, to some degree, with possible treatments of walls and ceilings.

The reason for inclusion of this chapter is mainly to state that the limits of your imagination are the only limits to the possibilities of wall and ceiling surface possibilities.

We do, however, feel that there are certain types of finish materials and applications that, for one reason or another, are not feasible. Our experience might save you some trouble but we *do not* want to discourage use of your imagination or experimentation with the unusual.

First, we will state that we have a certain bias for clean, white walls and ceilings in most areas because they help create a clean space which has the feel of being larger than it is. And wall-hung art is at its best displayed on a plain, light background, as your museum curator will testify. Sheetrock is a reasonably good, fairly inexpensive material to achieve this smooth surface and will be discussed below.

In some cases, however, the wall or ceiling, or a portion thereof, can *be* the art. In other cases, there will be no art because the resident wants none or because the whole is the artistic expression. A wall of stainless-steel plates pop-riveted to the backup in a High Tech pad would probably not be an appropriate place for a reproduction of Whistler's Mother.

One thing to consider is that whatever the interior layer, or cover, it is advisable to have a solid, continuous layer of something behind it to separate you from the wall insulation, or wall cavity, to impede the flow of air currents, dirt and vermin.

This brings us back to the possibility of sheetrock, but plywood is better if the finish material must be nailed up at points other than where the framing members are.

We have not talked much about sheetrock elsewhere in the book because it is not what we consider to be a material that can often be scrounged used or damaged. Sometimes edge-damaged stock is available but not at a greatly reduced price because almost every sheet is cut before it's nailed up anyway.

Sheetrock, gypsum board and drywall are all names for the same stuff. It is most often ½″ thick and 4 feet wide. Common lengths available are eight, ten and twelve feet. It is common practice to nail sheetrock to walls and ceilings with lots of nails or install it with mastic and much fewer nails.

Installation of sheetrock is fairly simple because it is the only material we have which requires that the installer be sloppy with his work. Joints should not fit tightly and after nails are pounded in flush with the face they need another hard blow to recess them into an ugly hammerhead hole which will hold finishing compound (commonly called sheetrock mud).

While hanging sheetrock requires little skill, finishing requires considerable. Joints and inside corners must be "taped," while outside corners require the installation of metal corners, which, in turn, require filling with mud and sanding, at least twice.

Finishing products and methods have improved dramatically over the past few decades and there are now pre-mixed mud and little hand dispensers for taping. We suggest watching some professional drywallers and acquiring some good equipment before embarking on your sheetrock project. Don't expect too much from yourself at first — in terms of speed or quality — but a little experience will teach you a great deal.

We have mentioned a few products which can be purchased new for prices we consider to be within reason. Sheetrock is one of these, along with cabinet door magnetic catches, pre-finished celotex board, marlite panels, polyethylene sheet plastic, construction adhesives, caulking, garage doors (installed by a manufacturer's authorized dealer), electric garage door openers, joist-hangers, aluminum storm windows, most glass and all insulation — because it does so much to save money over the years to come.

Before sheetrock there was plaster — and there still is plaster. Much more expensive and difficult to work than sheetrock, it is also better. And it is quite necessary to use when refurbishing an old house with areas of loose or missing old plaster. However, even there, it might be more feasible to remove the old plaster and replace it with something or install something like sheetrock or paneling over it.

Patch plaster and spackling compound can be used on smaller areas with great results, but they get expensive and difficult to manage on larger areas.

The degree of smoothness required on a sheetrock or plaster surface depends on the finish to be applied, of course. A surface to be covered completely with bamboo slats will need to be free-of-holes only. Tape over holes in sheetrock would be enough. If a heavy textured cloth or vinyl-wall-covering is to be used, the surface needs to be only moderately smooth, etc.

Surfaces to be painted are a different breed. And paints vary extremely in their covering and flaw-hiding abilities.

Two rules-of-thumb can be followed when looking for paint to camouflage less-than-perfect surfaces: (1) the darker the color, the less imperfection will show through; (2) the less shininess, or gloss, the less noticeable irregularities will be. High-gloss white enamel will allow the eye to zero in on every little not-quite-full nail hole or sanding scratch. Flat black latex will emphasize the whole heavy area and hide even offsets and gouges.

Unless your project is High Tech or includes a dark room, you will probably not want to go to either extreme. But if these rules are kept in mind along with condition of surface, size of the area and amount of abuse to which it will be subjected, a color-texture-finish paint combination can be chosen to best suit the condition.

Keep in mind, with paint, paneling or any other material, that dark color closes in on the viewer (makes an area seem smaller) and light color does the opposite. If a high ceiling of old, bumpy plaster in a small room is painted a dark, flat color, the bad plaster will be seen less and the ceiling will appear lower.

Another tip about color-choosing from paint books or material samples: the small piece of paint or material will seem much lighter than it would in a large area of wall, ceiling or drapery. An entire painted wall will be *several times darker* than the sample led you to expect!

Adding texture in with paint (like sand), or on the surface behind the paint, is a very effective way to camouflage surface irregularities. Your paint dealer can help you with additives.

Most sheetrock ceilings are textured today, because it is so very difficult to obtain a smooth enough painted surface for even good sheetrock work not to read-through when exposed to artificial lighting. The difference in texture between the paper field of the drywall and the slick, sanded areas of compound (to say nothing of the paper areas which were inadvertently sanded) are virtually impossible to paint in such a way that they won't "read" as a pattern.

There are, now, prefinished texture materials used mostly in ceilings, which are sprayed on, trowel applied, sponge-textured, brushed or rolled. Other, similar products require painting later.

Most wood, used inside, darkens with age. So does much wall-paper. Many old knotty-pine rooms, for instance, have "closed in" on residents who have grown older with the darkening process. It is good to guard against this phenomenon.

Hardly any materials are maintenance-free and some, we think, are so grossly misunderstood in this regard, as to be ridiculous. Ceramic tile, for instance, requires constant care and eventual re-grouting to be decent. Brick fireplace fronts are almost impossible to clean when they get smoke covered — especially the rough-surfaced varieties. Rough sawn wood gets rain-washed when used as exterior siding. Something else happens when it is wiped with a soft cloth in interior housekeeping.

Several interesting examples of unusual surface treatment which we have seen are worthy of mention along with comments on installation and results.

When this writer was a school kid working part-time in an architectural office, he was involved with some of the other young men in the office in a get-rich-quick project that didn't exactly work out. Well, the project worked out okay, but we didn't get rich. We bought an old house on Queen Anne Hill in Seattle, remodeled, repaired and sold it.

We had come into possession, somehow, of what seemed like an inexhaustible supply of folded paper hand towels like those that come out of some restroom towel dispensers. We spent a number of hours you wouldn't believe and an equally unbelievable amount of paste covering a rather large, dominate area of wall in a random pattern of towel-on-towel. There were too many pasters with too many ideas about installation, areas of too-much and too-little paste and too few towels to finish the job.

This writer's sister, Anne Treadway, stapled corrugated paper to the rather high and deteriorating plaster ceiling of her dining room — then painted it dark red. The paper was cut into slightly irregularly shaped pieces about 36″ square. It was not glued, to keep the weight from delaminating the old wall paper further, so there have been a few extra staples needed over the years. Otherwise, this project has been a total success.

An old residence in Santa Fe was remodeled and made into a restaurant some years ago. The ceiling just inside the entry door was quite low and visually dominant. It was, somehow, covered with white chicken feathers! Does anyone know what happened to it?

A friend bought gobs of bamboo-slat roll-up curtains at garage sales over a period of several months. They were in varying degrees of wholeness (broken ends, missing slats, raveled strings, etc.). After removing the bad areas, she installed them on the walls of her den with mastic. Before the mastic was too dry, she cut and pulled out the strings. Vertical lines between areas of horizontal slats were placed with so much care that the result is a really classy room.

And some friends of this writer, in Germany, suspended a piece of mirror out away from a bad place in a wall in making an area of a bombed-out building livable after WWII. During a party, a guest walked into it and broke it. Fortunately, there was very little physical damage to the guest.

Another friend has papered the walls of the bathroom in the guest cottage with pages from women's magazines. This type of treatment can be really cornball, but done with finesse and good craftsmanship, all those super-thin women in extremely abbreviated clothing, and with their too-pretty male companions and perfume bottles, are really very chic.

Carpet has been used on walls and ceilings with varying degrees of success. Like vinyl-wall-covering, sisal sheet material and heavily textured cloths, such as burlap, carpet will cover a lot of uglies while adding softness and warmth.

We think that using the same carpet on walls as the floor can be a mistake, because the un-trampled-upon wall carpet will continue to be so clean and fluffy that it will make the floor look soiled and ugly before its time. And heavy carpet is sometimes difficult to keep up, especially on ceilings.

Carpet does have some insulating value as well as being warm to touch and sight.

As often stated, insulation can't be over emphasized or overdone — so the "R" value can be a factor in the choice of a material for finish or backup for outside walls and ceilings.

One more possibility for sealing off a wall or ceiling inside, whatever your choice of finish material, which might not be so air-tight: poly-plastic film. It is needed for vapor-barrier on surfaces to the exterior anyway.

Another type of insulation which can be important has to do with sound transmission. Common sense tells us that softer materials absorb sound while hard, smooth stuff allows noise to bounce around and transmit through.

With all the uncountable possibilities for wall and ceiling coverings, only a tiny few have been mentioned in this and the other chapters. We would welcome descriptions and/or pictures of your ideas.

■

CHAPTER TWENTY

Floor Coverings

Because sheet-vinyl, floor tile, linoleum and glue-down carpets are not salvageable (they can't be removed intact), we will concern ourselves mainly with floor covering items which can be bought new.

The exception to this is used carpet and pad which was installed with carpet tack-strips.

This type of carpet is generally of superior quality to the glue-down type and is constantly being taken out and replaced in remodeled commercial buildings and offices. Fortunately for us scroungers, many companies moving to different locations feel compelled to hire decorators and change everything, including expensive carpets.

Except for being dirty and having holes cut for floor receptacles for electricity and telephones, this is often extremely good stuff. Most towns have companies which specialize in carpet cleaning and repairing. Take bids on having your carpet cleaned, the bad places cut out and pieces sewed together to form the right sizes for your various rooms.

Or you can buy some carpet tape, a hook knife, needles and thread, and save the expense. You will also need to buy new carpet strips to nail down around the perimeter of your areas to be carpeted.

Carpet stretchers and cleaning equipment are too expensive to buy for use on one project and can be rented. Go watch some carpet installers to see how it's done. Be sure your carpet is stretched really tightly. Otherwise it will expand and wrinkle in hot weather.

If you decide to glue your carpet down, do it with a solid coating of mastic. If you install it with only lines or spots of mastic, it will expand when hot weather comes and you will have hills and valleys on your floor.

■

As with all other manufactured products, some rejects develop in the manufacture of floor coverings. Some of these mismanufactured goods are sold at factory stores directly to consumers. Others are sold by the producer to salvage/surplus stores and floor covering dealers who specialize in rather low-end and "seconds" merchandise. These companies evidently pay near nothing for some of this stuff, judging by some of the prices we have paid.

There are, of course, varying degrees of "problems" which cause a product to be a "reject" or "second." Sometimes the quality is perfect but the color is off enough from normal that that roll, piece or lot could not be used with standard stock. On the other end of the spectrum would be a piece of carpet that got hung up in the machinery and came out so shredded that it could only be used as packing.

The people selling these products are usually among the least imaginative or creative in the world. Their contracts with the plants stipulate that they will take what is sent them. For this reason, the more off-beat stuff can often be obtained for the going price of packing material. As an example of this, the photo below shows some sculptured-pile carpet which got caught in the machinery when the color dyes were being changed. Some pretty wild streaks and spots developed which created a piece of art that could be entitled "African Safari." There were thirty-four square yards in this piece and it was sold for thirty-four bucks. Yes: one dollar a yard!

town to see what's available. And ask to see what is to be thrown out. Remember that sellers of first-quality, new merchandise often fail to see any value whatsoever in what might be, to you, the most perfect solution to a design problem. At carpet outlets ask to see the stuff set aside to be sold for packing.

Keep in mind that materials manufactured for floor covering don't necessarily have to be used on floors. And, by the same token, no law says that you must cover a floor with a product called a "floor covering."

$1-a-yard carpet which had become stuck in the dye-machine at the factory. In the dome bedroom/bath of author's house.

Wholesalers, retailers and installers all end up with discontinued stock, pieces and lots too small for normal use, etc. This stuff has to be moved out to make room for profit-producing merchandise and can often be obtained for a fraction of regular prices. Take a trip around

Carpeted walls, cabinet ends, doors, even ceilings, can be attractive and suppress noise. And carpet is surprisingly good thermal insulation — carpet and pad have a combined value of about R2 (compared to that of vinyl sheets or tile of about R.05).

We have seen many instances of carpet sample pieces being pieced together on floors and walls. Some of these applications were "cute," a few were rather nice and most were ugly. Carpet samples are often available free and always available very cheap. If you do use them, we suggest that you collect enough stock to choose from to obtain compatibility, not only in colors but in types, textures, thicknesses and content, as well. Not all carpets clean the same way, for instance.

Piecing together small pieces of sheet-vinyl or linoleum usually results in disaster, because it is virtually impossible to make cuts that will fit tightly enough.

Resilient floor tiles (vinyl, vinyl/composition, etc.) from odd color lots and designs can be used interestingly and practically, if they are all the same thickness, but there will be enough difference in size, from one lot to the next, that it might be necessary to lay them in stripes or irregular patterns.

Ceramic and clay tiles are often available in rejects, odd lots, etc. This is much more difficult stuff to cut and work with than resilient materials, but the finished product can be well worth the effort.

This "hard" tile can be set in a bed of mortar or installed with mastics, which were developed for this use, then grouted and cleaned.

Cutting requires a tile saw, which you might be able to borrow from the store where you were able to find your treasures. Or the store and rental companies will have them for rent. These saws use blades made with abrasive chips and use a flow of water over the blade and material being cut. They are not hard to use and, unlike most other equipment, are not dangerous, since the blade will not cut skin on contact. But cutting and fitting is slow.

Some real works of art have been achieved on floors, walls, cabinet tops, backsplashes, recessed tubs, etc., with the use of odd-lot tiles. We would appreciate being sent a picture of your masterpiece.

■

Stone, brick, marble and other hard-surface materials have been used effectively, in innumerable ways, for flooring. And dark stone or brick is the most desirable material for the exposed layer of a solar energy "sink," or floor collector mass, because it will absorb solar heat.

On the other hand, stone and masonry floors are cold and uncomfortable to touch in areas not exposed to sun rays. And the more irregular and porous the floor surface (with grout lines, etc.), the more difficult it is to keep clean.

The photo on page 116 was taken of a south-facing room with a floor covered with broken pieces of blackboard slate which the owner was allowed to take, free, from an old school building which was being remodeled for a different use.

Please use non-combustible material for flooring for at least thirty inches in front of fireplaces and below and in front of wood or coal stoves.

Rugs and area carpets can be beautiful over stone and brick floor during summer months.

■

Prefinished wood flooring in strips, blocks and parquet tiles is sometimes available to the scrounger. This stuff is easy to install with nails and/or mastic, and blemishes are not usually too difficult to repair.

We have covered the use of other wood products as flooring in previous chapters.

■

Covering "on-grade" or "below-grade" concrete floors with finish flooring materials requires extra thought and care.

When a concrete slab is poured it should always have a moisture barrier below it (see the chapter on "Insulation, Vapor Barriers and Sealants"). But no membrane is absolutely safe protection.

Broken slate, from salvaged blackboards, used as flooring.

There are special mastics for installation of vinyls, tiles and carpet on below-grade slabs, but they won't work as protection against "hydrostatic pressure," or moisture coming up through the concrete.

Wood flooring can be installed over concrete to help eliminate the basement "feel" and, when insulation is used, the floor can be made much warmer. But there is only one sure-fire way to do it which will not result in the wood buckling up and being ruined. This involves laying "screeds" over a moisture barrier such as polyethylene plastic sheeting and fastening your lumber or plywood flooring to the screeds.

Screeds are pieces of wood laid perpendicular to the direction on the flooring, between the flooring and the concrete slab. They can be of about any size which will allow the fastening of the floor and they can be installed loose, with construction adhesive, or nailed with concrete nails (power nailers are great). Most flooring should not span more than about 14″ between screeds.

Installation of poly-plastic or felt-paper between other types of subflooring and finish-flooring is often advisable too, to reduce infiltration of air, moisture and dirt and to retard noise.

■

CHAPTER TWENTY ONE

Furniture and Art

Government-auction items in Los Arcos Mexican restaurant.

The style and "feel" of your decor need only be governed by your taste. If you can't afford a new Barcelona seating group, an eighteenth century French square-grand piano, or a Renoir original (and who can?), you *can* create the "feel" you desire with little expenditure of money — perhaps a large expenditure of time and energy, depending on what you want to

achieve and your luck in the search — but not much money.

The most unbelievable objects get tossed into the trash and into the scrap metal heap for equally unbelievable reasons.

Pieces with broken legs get pitched because having legs made is usually not economically feasible, when a box-base could be substituted for the legs resulting in a modernized, greatly improved design. In this case, the base could probably be painted a dark, receding color, so the wood could be very common. And perfectly good table bases get tossed out when the top gets damaged.

Sometimes, of course, perfectly good pieces are disposed of, for one reason or another, for example, the $270 lounge chairs Henry Pina picked up at the Albuquerque Hilton Hotel for $40 and the 110 restaurant chairs he got at an Air Force auction for $22 each. At the same auction, he paid $110 for a 72″ TV!

We have found reupholstering to be terribly expensive to have done and very time consuming and demanding to do properly oneself. Slipcovers, on the other hand, can often be used beautifully at greatly reduced cost.

Stripping and refinishing of ornate pieces is also tough and expensive to have done by professionals. But, to some, a refinished old piece is worth about any expenditure.

We do not happen to be antique fans and are even less drawn to period reproductions, but we certainly realize there are those who do appreciate those things. We see good antiques being purchased for thousands of dollars and well crafted reproductions selling for more than contemporary pieces (as well they should, because they are more difficult to produce) and we hope they are used sensibly and with taste.

As with stained glass, antique hardware, ornate marble fireplace fronts and detailed millwork, a piece of period furniture is going to look good if it is compatible with the surroundings. No matter how good the pieces, an empire table, Victorian highback chairs, an early-American hutch and a Danish-modern serving cart are all going to look like sin-in-the-morning if they are used together.

On the other hand, some people can manage mixtures of some periods in good eclectic design (it's tough to do), and old pieces can sometimes be spotted around in a more modern interior with stunningly beautiful results.

Furniture and art do not, of course, need to be conventional four-legged tables and chairs and rectangular, framed pictures hanging on walls.

The bed in the dome upstairs bedroom/bath of this writer's house is a 4″ foam mattress on a panel made of two damaged solid core doors. This panel is on four rubber-wheel castors that came off a store fixture bought at an auction for $2. A 1 x 8 pine skirt is held off the floor about ½″ and the bed rolls around on upright pipe-with-flange which was obtained at the junkyard and was originally used for who-knows-what. This allows the bed to be turned to take advantage of whatever the heavens are best displaying through the plexiglas roof that evening.

At the opposite end of the spectrum from antiques would be either "Stark" decor or "High Tech." Some designers have actually won recognition by "furnishing" a room with a carpet to rest on (presumably) and one piece of grotesque furniture, usually a chair, to talk about (evidently) while resting.

High Tech, we feel, has considerably more merit. Stainless steel, glass, plastic, chrome and enamel combine to form every utilitarian piece of furniture, as well as every object which functions only as "decoration," with no frills or softness.

To assemble a High Tech interior, one needs to be on the lookout for closeout sales in such places as restaurant supply houses, metal fabricating companies, medical supply firms, and commercial/industrial plumbing companies. Junk yards and scrap heaps yield an unending supply of absolute treasures which, with some gloss enamel, will become stools, benches, table bases, shelving units, bed frames, hall trees, exercise equipment, towel racks, lighting fixtures, standing and hanging art, etc.

Some items of commercial hardware, such as panic-bar units for exit doors, can be picked up at salvage sites along with big clothes hooks, hospital curtain tracks, shower seats, and grab-bars.

Most High Tech we have seen has been done very coldly, with stainless steel and white, for the stark effect. Bright color enamels can produce a somewhat less high High Tech which can be more fun and, for many, much easier to live with.

Straight-line contemporary furniture is, in our opinion, the easiest to design, build, cover and finish to produce inviting, comfortable living. This stuff can vary from very stark, almost High Tech, like seating in airport terminals, to softer, warmer, more round-edged pieces.

It is a solid-core ash door (damaged beyond hope on the bottom side) hung from the 3 x 10 roof/ceiling beams with a steel assembly fabricated out of scrap steel at a junk yard. The chairs are discount-store sale specials and the enamel color on the "hanger" was chosen to match the chairs.

The hanging "ball" light shades over the table were made by wrapping strips of the inner-bark of a birch tree, dried, around wire frames. Coat hanger wire is the perfect gauge to bend into whatever shape you might want. Manila twine pulled apart makes good wrapping, as

Dining table top (a damaged door) hung from ceiling by a scrap-metal hanger. Cheap chairs covered with carpet remnants.

Table tops of wood, formica, plastic, glass, marble, porcelain, plexiglas, metal, or anything, can be held up with easy-to-make square legs or pedestals, or they can be cantilevered from walls, hung with chain, etc. The photo above is of the dining area table in this writer's house.

would strips of plastic cloth or a zillion other things.

Built-in furniture has certain advantages over movable pieces, but is certainly not for the person who wants to keep changing the layout. And built-in beds in alcoves, corners, or even against a wall on one side, are pains to make up. But pieces hung from walls or ceilings, sans-legs, or on solid bases, make floor-cleaning easier, there is usually more stability with built-ins and they are, generally, more compact and space-saving.

Houses with built-ins, when put up for sale, are advertised "lots of built-ins" because people love them and will pay a disproportionately higher price for a house with them.

Shelving, magazine storage, etageres, etc., can be some of the least expensive, most fun and pleasant-to-behold features in any setting, whether replacing broken glass and repairing Ethan Allen stack-boxes in a traditional setting, assembling waterpipe racks with aluminum grate shelving in a High-Tech room, or hanging some shipping crate-boards on the wall with scrounged store fixture hardware.

No expensive or ugly shelf hardware.

Closet/desk/bookcase/storage units can become the dividers between rooms, eliminating the cost of a wall, saving space and affording some degree of flexibility of floorplan. And art alcoves and display areas can become integral parts of the units.

To many people, there is beauty in books on display far transcending the colorfulness of the covers, to say nothing of the joy of having treasures surrounding us in our castles.

One of our favorite ways to install shelving is shown in the photo above. The notched pieces of 2 x 4 support the shelf while furnishing partial bookends. The same thing can be done with separate pieces between shelves for uprights — which allows the shelves to fit back against the wall.

In custom furniture design, few rules need to be followed, *except* in seating. Tables and desks need to be 29½″ tall (26″ when "continental" height chairs are used), book shelving should be much less deep than one would imagine (standard library shelves are 6″ deep; paperbacks are about 4¼″ and standard hardbacks about 5½″), and that's about it.

But for something to be comfortable for an average American-type person to sit on, it must follow several quite rigid design rules. The front edge of the seat should be right at 16″ high. The seat surface should be between 17″ and 20″ deep and *it must slope down to the back* at least an inch and not more than 2″. If the seat is too shallow it will seem like a ledge; if too deep a short legged person's knees won't bend if his back touches the bolster, and if the seat does not slope down to the rear, the sitter will slide out the front, which is a most uncomfortable sensation.

The back, or bolster, must also be slanted-away from the seat at the top and it should make contact with the human back from about 3″ above the back of the seat to about 15″ above.

There are many comfortable chairs which break these rules — some of them drastically — but they have all been anatomically engineered, tested and redesigned and perfected. And, even then, many famous chairs, such as the Ames Mister chair, the Bertoia High Back and the Fritzhansen Egg chair, are very difficult to get out of for someone who is pregnant or has weak legs.

Use of foam in hand-made furniture is discussed in the "Plastics" chapter, as are several other materials in the various other chapters.

Beds can be comfortable in about any size, shape, or height, as long as no part of the occupant hangs off, but there are a few things to consider before spending time and money. Bed clothing is manufactured in a few standard shapes and sizes and fitted lower sheets sure are handy. Mattresses and box springs are also certain sizes.

About everyone has a favorite degree of softness or firmness and some require a certain type of mattress to avoid back problems. This writer prefers a good 4″ foam pad over a solid panel, but not all foams are the same. Good foam mattresses made by Goodyear and other large companies are specially molded for support and contour at the same time. Be careful buying foam mattresses at garage sales because what you are likely to buy, unknowingly, will be a mattress out of a recreational vehicle, too short for standard sheets and, for many people, comfort.

At the time of writing, a king-sized waterbed with base, frame, head-board, liner, heating element, thermostat control and mattress costs about $250. Everything except the base, frame and headboard can be found for about $100. We would strongly suggest buying a high-quality mattress with baffles (to slow the water movement — thus adding some firmness) and building your own frame with whatever accessories and in whatever style suits you.

Water is heavy, especially water in motion, so it's good to know that there is plenty of support structure below where a waterbed will sit, especially if any activity other than sleep is anticipated.

Furniture pieces are small enough that they can often be built of pieces of materials too small to incorporate into the house structure.

In 1954, this writer, at age 24, disassembled an old upright piano, which would not stay in tune, and used the parts to build 11 pieces of furniture and accessories. The May 1955 issue of the *American Home* magazine published an article with photos about this undertaking, which gives some indication of the years of my interest in creating with junk and other good stuff.

One fellow we know, who loves automobiles, hung bucket seats out of a wrecked sports car on a support post in his finished basement with pretty good results. We were less impressed with his bar made out of car body parts but he, and a lot of his party friends, think it's great — so it's great. His bar stools are tractor seats on axles and his walls contain interesting groups of hub-caps, hood ornaments and lighted instruments showing a speed of 92 mph and ¾ of a tank of fuel.

We own, among a gob of other scrounged treasures, a complete set of an upper and lower pullman car seat/bed unit. There are the two seats, which face each other and make into a bed, a pull-down bed unit with mahogany-veneered steel case and all the hardware, wrench-even sheets, and pillow cases to go with it (all embroidered with the name FRISCO which was the nickname of the St. Louis-San Francisco Railroad before it was merged into the Burlington-Northern system).

And we have a toilet compartment door, caboose lantern and various other railroad goodies. Someday we will hear of someone who is building and is a nut for trains and we'll install a really unique spare bedroom!

Bob DeLong, my scrounging partner, is also a drummer. Bob acquired, somewhere along the way, three old brass kettle drums of different sizes. When his building project is complete, these will be made into two planters, to hang below skylights in his very unusual greatroom, and a table with glass top.

Products like whiskey barrels, cable reels and concrete-block 1 x 12 shelving units have, perhaps, been overused, but there are innumerable less-used or never-before-used furniture makin's out there.

Wherever you are looking at whatever kinds of things, if you "think furniture," you will come up with a thousand more ideas than you have room to use.

Art, according to one of Mr. Webster's definitions, is: "Products of creative work." Using this definition, everything we are discussing here is art, whether it is communicating with Building Regulations Department people or shaping styrofoam packing moulds into cavity insulation.

Certainly any item created by man for the purpose of being viewed by man is art. Art and beauty are not one and the same and beauty is, indeed, in the eyes of the beholder.

An original Picasso line-drawing has tremendous monetary value and, to some, equal beauty, even if beauty was not what Picasso wanted to create. There are others who would really not want a silhouette of a person with two eyes on the same side of the head hanging in the living room.

On the other hand, what you might create to hang on your wall, or down from the ceiling, or to stand among the trees, or on a shelf, could well be beauty to you and most others who view it, while having very little intrinsic value.

To at least some degree, Louis Sullivan was correct when he stated that "form follows function." We would like for you to build a functional shelter which has individualistic form displaying your own personality and creativity. We would like for you to carry this ambition and creativity inside to form surroundings which give you the pleasure you deserve and that you can take deserved pride in.

Your entire product will be a work of art.

■

CHAPTER TWENTY TWO

Appliances

While most items which one might scrounge and piece together into a dwelling can be used somehow, effectively and safely, regardless of condition — appliances require more scrutiny.

The energy fed into an appliance, whether it be electricity, natural gas, bottle gas, coal or wood, is costly and potentially dangerous.

A free-for-the-taking water heater, range, or anything else is not a bargain if it endangers the lives of your family. Neither is it a bargain if energy pours through it inefficiently or if it will require costly repairs.

There are, however, several sources of perfectly good appliances other than the new appliance store. And even they, sometimes, have bargains in units which have been damaged or have set around long enough to have become outdated. An advantage is that these are often covered by guarantees.

Scratches and dents on appliances are usually where they won't show if pushed up against a cabinet or wall. Some units made to be free-standing can be built-in either with the housing left on or with part of it removed. Refrigerators require some study before being built-in and this will be discussed later.

■

Your water heater will probably be your appliance most deserving of research and, certainly, of care taken on installation and insu-lation. If you heat your water with fossil fuel, safe venting is the primary consideration. Like a furnace, heating stove or boiler, the water heater should be installed by someone who really knows how vents work. If you install it yourself, be sure you study first.

A surprisingly large percentage of utility use goes for water heating. Huge amounts of waste occur because we keep our water temperature controls set too high (so that we have to run cold water with the hot to keep from scalding ourselves), we don't keep our burners clean and there is never enough insulation covering our hot water tanks.

We strongly recommend the installation of a gob more insulation on the tank and pipe-wrapping on the hot water pipes. There are also automatic flue closures to use above gas-fired units. These are effective and not terribly expensive, but they *must* be installed correctly.

The only reason that a hot water heater should not be regulated down to a much lower temperature than most folks use is having a dishwasher that is not equipped with an inte-gral water heater. Many models are now equip-ped with these — most older (or cheaper) ones are not.

Solar energy domestic hot water systems are available or can be homemade by someone with knowledge and ability. And likewise, heat exchangers at water heaters can be attached to coils in fireplace units, over wood furnaces, etc. Care must be exercised to protect against water contamination as well as fire.

■

Old dishwashing machines are readily available just about everywhere, because they cost so much to operate and people are taking them out. We suggest you not buy one of these and, if you own one, that you not use the drying cycle — just let the dishes sit and dry slowly.

When you plan and build your kitchen cabinets, you can make a 24″ wide area next to your sink easily removable for a future dishwasher if you can't afford a good, efficient one at the time. Residential dishwashers are all made to fit into a 24″ wide by 34¼″ tall (floor-to-bottom-of-cabinet-top) opening. One-hundred-ten volt electricity will need to be provided. Adapting sink water and drain piping is not a big project.

A dishwasher can be hooked into single or double compartment sink plumbing, either with or without a garbage disposer.

■

If you do not want to use your garbage for gardening and especially if your plumbing will be hooked into a public sewer system, a garbage disposer can be used very reasonably. But know new prices before paying for an old one. New units are not terribly expensive and used ones usually require new cutters if they work at all.

Again, it is simple enough to install a garbage disposer at a later date, but another 110 volt electricity supply will be needed *plus* a switch, not always easy to find a place for and install after the project is complete. It is easy to put a box and some wire in with the rest of the electrical work.

In-Sink-E-Rater garbage disposers and Kitchen-Aide dishwashers, we think, are the tops in their fields.

■

Trash compactors are standard sized at 15″ x 34¼″. It is not likely that you will find one used yet or damaged unless the damage is too severe for hope.

■

Cooking stoves can be found in about any condition you are willing to accept. Most free-standing and built-in units with single ovens below are the standard 30″ width. It is not necessary to allow any more space than exactly what a range requires, because they don't have to "breathe" like a refrigerator.

Electric ranges generally cost more to operate than gas ones and most chefs prefer the regulatable flame of gas. Electric ranges found used are also very likely to require new rangetop and oven heating elements. On the other hand, electricity is cleaner than gas and in areas where natural gas is not available, it might not be worth the bother to have bottled gas just for heating water and cooking.

In buying any pre-used gas burning device it is imperative that it is determined what kind of gas the burner is manufactured to burn — natural or bottled. Burners can sometimes be modified or changed, but the cost and bother might not be worth it.

Separate rangetop and oven units are available in both gas and electric. There is no standardization of size of these units, so find them before building your cabinets. Two-hundred-twenty volt electricity must be run to each cooking unit with separate circuits. A combination range-and-oven requires only one 220 outlet—rangetops and ovens require separate outlets (and circuits) if they are not combined.

There are certain old units which you might find that you would be well advised to buy if the seller doesn't realize what he has.

Old Thermadore units are all superior and anything with a Chambers nametag is *really* a find. Chambers ovens are extremely well insulated and have the best burners and controls ever installed on American appliances.

Sometimes warming ovens are available at sales and salvage stores. Be sure you don't buy one thinking it to be a baking oven.

■

Microwave ovens are also sometimes available. We do not recommend these being purchased used or at any kind of a "bargain" deal. Study microwave ovens before spending your dollars.

■

Wood-burning antique ranges are expensive when they can be found in good enough shape to use. New units are being manufactured (also expensive) which have much better insulation than the antiques and are much easier to regulate.

Combination heating-cooking woodstoves are also available as are in-flue ovens, fireplace cooking items, etc.

Farm auctions sometimes include wood-burning appliances, but they don't usually go cheap.

■

Refrigerators are the most complex of our modern kitchen appliances and deserve considerable consideration. It is an amazing thing that there are still many G.E. and Frigidaire boxes from the twenties and thirties still functioning very well while there are twelve-year-old units of the same brands which are worn out past the point of repair being economically feasible.

Compressors often go bad on relatively late model refrigerators and that's it. (For $5 the guy that brings your new refrigerator will haul away the old one.)

And self-defrost mechanisms are *terrible*, repair-wise. Even door gaskets for a box built in the sixties can sometimes cost more to replace than another old refrigerator, whole.

Not all used refrigerators will go bad the week after they are restarted — but there is not likely to be a guarantee.

Freight damaged refrigerators often show up at "salvage" type stores because they are so large and awkward to handle. Banged-up sides, tops, and even doors can be fixed with panels of thin plywood, cork, vinyl wall covering, formica, marlite, or whatever has been scrounged. But "building-in" a refrigerator to hide its damage or age requires planning.

Refrigerators exhaust the heat which they extract from the space cooled. Some exhaust at the rear and others from grilles at the bottom-front. And for any air exhausted some make-up air must come from somewhere to replace it. We have seen some clever trap-door devices and other interesting ideas implemented to allow the hot air exhaust to blow outside in the summer and for cold air to come from the exterior in the winter. Care must be taken to not get the reverse action going and to keep from burning the unit up.

And doors through exterior walls which are not well insulated can cost considerable in fuel waste.

■

Freezers are not essentially different from refrigerators in terms of where they are located or built-in.

Used freezers are often available at very low cost. Your newspaper ad section will usually have several listed. Some of these, like chest-type units from the forties and fifties, are real bargains. And old ice cream boxes with several lids on top are sometimes available. These ice cream boxes are usually very well insulated because ice cream must be kept colder than most frozen foods to be hard.

But be certain that a newly acquired freezer functions well before putting a lot of expensive stuff in. It would take a lot of conservation to make up for the loss of a freezer full of food.

■

Clothes washers and dryers are always available used, damaged, at "close-out" and at all kinds of sales.

Washers are fairly complex machines which can require costly repairs after several years of hard use. They have also become more versatile over the years with the newer models having more "modes" of cycle length, spinning, and agitating speeds, temperatures in wash and rinse phases, etc.

Some of these features might appeal to you and anything that will save energy is worthy of consideration.

Dryers, too, are getting more sophisticated and complicated, but they are much more simple than clothes washers. Consequently, they are much less costly to have repaired and they are usually fairly easy for the handyman to work on.

In areas which have natural gas, gas dryers are considerably more efficient (thus less costly to operate) than electric ones.

There are provisions to be made during construction on a project which will include a washer and dryer later when money is available. Hot and cold water pipes with hose bibbs and a drain pipe are needed for the washer along with 110 volt electricity. An appliance store will give you more detailed requirements.

Dryers require either gas or 220 volt electricity from a separate circuit. Dryers also require venting which requires some planning. A 4″ diameter vent pipe, or hose, must connect the rear of the machine to a place where lint and a tremendous amount of hot moisture can be dumped without causing harm. Venting into a room is harmful to the structure and terribly inefficient. Exhausting into a crawl space below a house is not much better. And there is certainly a limit to how far and through how many turns and bends this steamy mess will travel.

It is highly desirable to have a dryer located on an exterior wall or very close to one.

Of the older model washers and dryers, the better quality ones might be Whirlpool and Speed Queen.

■

With appliances, as with the rest of your project, the two watchwords should be safety and fuel economy.

CHAPTER TWENTY THREE

Electricity and Lighting

The former is a very exact discipline; the latter is not.

Because of the exactness, and the danger in straying off the exact path, there are very rigid rules set down in codes and utility company requirements governing electrification of buildings. Not just commercial buildings, but projects such as yours also.

Generally speaking, it is unlawful and unwise to incorporate pre-used or sub-standard electrical equipment and apparatus into new construction. There are a few exceptions.

While it is unlawful and foolhardy to allow electrical work to be done by someone who does not understand what he is doing, it is not difficult to learn how to do basic house wiring. The library has books, as do trade schools and electrical companies. The law says that a person may do his own electrical work if the building he is wiring will be used solely by that person and his or her own family.

Learning electrical work from a textbook will not, of course, make you a master electrician. Your work will be slow going, especially at first, but there is something very rewarding in "checking off" each outlet which has been connected and each "home-run" completed to the main box from your list or plan.

Unless you are very comfortable working with electricity, it would be wise to have a good electrician or engineer check your system over before the "moment-of-truth" when your utility company connects it to its supply.

Except for light fixtures, and materials to use for fixtures, there is not ordinarily much electrical material to be salvaged from demolition sites. It is extremely rare to find a circuit-breaker box or fuse-box which would be good enough to reuse even if allowed by the code. If you do stumble onto a *good* breaker box, however, which is a reasonable size and in good condition, grab it — you can probably obtain permission to use it.

Disconnect boxes are sometimes reusable and they are extremely expensive to buy new. The same is true of "mastheads" (the pipe which sticks up above the roof that the service comes in to).

Junction boxes, switch boxes and receptacle boxes are often salvageable. These are not such expensive items, but it all helps, certainly.

Wiring which was pre-used, exposed (not run in conduit), is not salvageable for reuse. It is worth something, stripped of its shielding, at the junk yard — depending on the going rate for copper — but it must not be reused to conduct electricity. The insulating sheathing will have disintegrated to the point that rebending, etc., in its removal will further damage it and render it hazardous and unusable.

Old switches and receptacles are a poor risk for reuse. Switches get loose and wear out. Receptacles burn out and disintegrate. And older receptacles are not grounded (they do not have the round holes for the ground prongs of modern plugs) so are illegal for use.

Conduit is often available where commercial and industrial buildings are being removed or renovated. Often, too, the removed/discarded conduit has wire in it. This wire *is* generally reusable in conduit because it has been protected and it tends to be of large enough gauge and in long enough lengths to be utilizable in a residential building project.

The codes or utilities regulations governing work in your area probably do not require residential wiring to be run in conduit, except in basements or crawl spaces. On the other hand, it is a better, safer way, and if you latch onto a bunch for free, it would be worth the extra effort to use it.

The reason conduit is often discarded is because it has been cut up and bent so that it is not economically feasible to pay a union wage to rebend and splice it.

When it is found, it will have all sorts of junction boxes and fixtures hanging on the ends.

It goes without saying that no splicing of wire should ever occur except inside of a box. It is also very important that all boxes should have cover-plates. So when you are picking up old boxes, try to locate the plates, too.

While you're at it, you can pick up all the wire-nuts lying around loose. And if you decide to use conduit, try to save the box connectors and nuts.

In any case, you will probably need some conduit (large size) for service to your breaker box. And various-sized conduit can be well utilized for all sorts of non-electrical uses from closet rods to furniture legs.

Sources of electrical materials, other than from demolition sites, are surplus stores, auctions and close-out sales. Caution — know your requirements and costs of items for sale at the supply house before going on a buying spree.

■

Lighting can be fun! It's fun to place light where it will do the most good, set the best moods, and create the nicest effects.

Light fixtures of extremely diverse varieties are to be found at demolition sites. The photo on page 129 is of a pile of old 16″ diameter green and white industrial lenses from a railway freight building. Several of these are perfect and they have conduit that might be used, as is, for hanger-arms. Some eight-foot, four-bulb fluorescent fixtures also came out of this building intact, as did some odd wall and ceiling mounted globe-type fixtures (with broken globes).

Old white-porcelain receptacles are to be found, in abundance, at salvage sites. Some have integral plug-ins and others have pull-chains. The latter hardly ever work anymore.

Demolition sites and materials-salvage yards are just two types of places where light-fixtures, and things to make light fixtures out of, are to be found. At any place else in the world where items can be found there are potential light reflectors, shades, brackets, hangers, shields, diffusers, lenses, poles, stands, etc.

In many of the previous chapters, we have mentioned possible uses of materials in lighting — because practically any material can be used to good advantage this way when coupled with some ingenuity.

We mentioned the use of aluminum pie plates for shields in a church by the late architect Bruce Goff.

Recently we used some one-pound coffee cans for recessed ceiling "canlights." We had picked up some fixtures at a church white-elephant sale each of which had eight small porcelain receptacles on the ends of what had been ⅜″ diameter white cord. These fixtures had previously hung in a Holiday Inn dining room where, over the years, they had become so caked with grease and dirt that no amount of cleaning would rescue the cord. Consequently, we bought these fixtures for a dollar each.

The porcelain receptacles had stems, washers and nuts which fit a ⅝″ hole drilled in the can. We drilled a half-dozen other holes for heat to escape. Scrap sheet metal was fashioned into snap-on cover plates, or "rims," which beam the light where we want it — in this case around the perimeter of a parlor-grand piano from a seven-foot high ceiling. We still have lots of receptacles for future projects.

If you are not already aware of the vast selection of available light bulbs, it would be advisable to visit a good supply house and take a look before the lighting stage of your project begins. Many bulbs, from Christmas tree to 5 inch "ball," can be the fixture. Others will allow

design of fixtures of unusual shapes and sizes to create almost any desired effect.

As rigid rules must be followed in wiring to assure safety, logic must be exercised in creating lighting. Bulbs generate tremendous heat (the higher the wattage the hotter), heated air rises, metal conducts electricity, sharp edges cut into wiring and any material except asbestos can melt and/or ignite. People know these things — the knowledge must be applied.

Industrial light fixtures removed from razed R.R. freight building.

CHAPTER TWENTY FOUR

Plumbing

No law that we are aware of can prohibit you from being your own plumbing contractor and doing the work yourself, on a building project for your family. This might save you considerable money, but there are a few situations of which you will need to be aware:

1. Building in an area in which a plumbing code has been adopted, you will be required to have your plumbing system inspected and approved by the building regulation department before hookup to sewer, water and gas.

2. In a rural area not governed by codes you will probably need to have your sewer, water and gas piping inspected and approved by the utilities company (or companies) before tying into these services.

3. Some codes and utilities companies prohibit the use of used materials altogether, and some have lists of forbidden uses of salvaged pipes, etc.

4. Much of the used pipe and fittings available will be too rusted and filled with deposits to be reused as plumbing without extensive cleaning.

Because plastic piping has now become so extensively used, you will probably find lots of the old, less-in-demand, galvanized and black steel pipe and fittings around demolition sites. You might also find some plastic and copper stuff.

Steel pipe is often worth accumulating for uses other than plumbing, such as clothes rods, legs for furniture and work benches, room dividers, towel racks, lamp tubes, etc. And pieces of pipe held out a half inch or so from the face of doors, cabinet doors and drawers can become very fine, bold and classy handles.

Sometimes pipe is taken out, at a remodeling project or from a house or building being removed, which is good enough, with a little cleaning, for reuse as plumbing. Use copper, galvanized and plastic for hot and cold water, galvanized and plastic for sewers and vents, and black pipe for gas. The use of copper for gas is dangerous over the long run, and if you find copper large enough for drain pipes it would make some sense to polish it and use it decoratively or sell it at the junk yard.

Sinks and lavatories are often plumbed with chromed "tin" drain pipes and P-traps, and copper supply water kits. If you find these in good enough condition to use, be sure to keep them. They are ridiculously expensive to buy new.

It is, of course, better to determine how pipe was used before running your drinking water through it.

For portions of your plumbing system for which you have no stock of accumulated material, we strongly suggest you use new plastic. The reasons are simple: it is less expensive and much, much easier to work with. Be sure, however, that you buy the right stuff, of proper size, for each job you are doing. Gas, for instance, requires special plastic and it also calls for use of larger pipe than water does, usually not less than one inch inside diameter (bottled gas piping can be somewhat smaller).

Plastic is manufactured in several different types and about any diameter. And all necessary fittings (bends, branches, increasers, elbows, tees, cross-tees, P-traps, couplings, unions, caps, plugs and flanges) are available.

Perforated plastic pipes are made for lateral-fields (see below) and foundation perimeter drains.

If you use steel and/or copper, the same things are available to fill out the missing pieces. Do not join steel and copper together, directly, below ground.

It will save immeasurable time to draw out your entire systems of hot water, cold water, sewer (with vents) and gas, showing every fitting needed at every turn, etc. Then make an inventory of what you have, and acquire the needed parts before you start working.

■

If you are building in an area with a sanitary sewer to hook onto, you will probably need to get an inspection and pay a connector fee. If there is no sewer where you build, you will need to devise your own.

The most common private system is the septic tank with "lateral fields." This is a fairly inexpensive and easy-to-install bunch of stuff, but codes adopted by some areas severely restrict septic tank use.

In our home city of Springfield, Missouri, for instance, no septic tanks may be installed within the city limits. In our county of Greene, it is not permissible to build on any piece of ground less than five acres in size, except in subdivisions which have approved public sewerage systems. And even on a five-acre or larger tract, a system design must be approved after a "percolation test" has been run by a registered, professional engineer. These are probably near the most restrictive of the nation's regulations, but they are not uncommon.

On the other hand, most of the counties around ours have either much less restrictive laws governing waste disposal or none at all. This is curious when you consider that there are so many rivers and lakes in these counties, but enforcement of regulations would be virtually impossible in the remote areas.

Manufactured septic tanks are usually concrete, but there are some made of fiberglass and metal. It is possible to make your own, where the law allows, out of concrete, masonry or salvaged metal, like an old boiler tank. Your library will have a book with plans and recommended capacities for various sizes of families. Generally, a tank of five-hundred-gallon capacity is minimum for a hermit, while eight-hundred gallons is more standard for a small family and a thousand-gallon tank would accommodate a group of six or so. The tank is buried only a few feet from the house.

Diaper washing and using a garbage disposer both put a lot of strain on a septic tank.

A lateral field can consist of one, two or several pipes with holes in them which run out from the septic tank about 18″ below the surface of the ground (depending on area freeze depth). Inexpensive plastic pipe is the easiest, and probably best, lateral line material made. It is generally best to dig a trench for your pipe, deeper than the pipe will lie by six inches or more and about a foot wider than the diameter of the pipe, which is usually four inches, with gravel installed in the trench to surround the pipe. Cover the top few inches with dirt and seed over it. A continuous drop in grade from the tank to the end of the field pipe is, of course, a must; but don't let it drop more than two inches per hundred feet. Sometimes this requires curved runs which follow the contour of your yard.

Draw a diagram showing the exact location of the clean-out door on top of the tank (from corner of house) and keep it taped up somewhere. Even the best installation will require being pumped out eventually.

The number of feet of pipe in your lateral field will be determined by running a "percolation test" (or series of them where soil type may vary on the lot). This test involves digging a hole in dry, unfrozen dirt, pouring a given amount of water in and timing how long it takes for the water to "percolate" away like hot water through ground coffee in a coffee maker.

Again, your library will have a book on testing procedures with charts showing depth of pipe for your area, amount of pipe needed based on tank size and test results, etc.

Or you can probably be nice to a civil engineer and get him to tell you what you need or show you his chart.

If you are building a group of dwellings in a rural area you might want to consider one system, such as a sewage lagoon, to take care of the group. Different climatic conditions, regulations, soil conditions, area size, etc., will govern what you should do. If one of the group is sharp with this sort of thing, you might be able to research your situation and design a system you can build with the materials you can find and your own labor. Otherwise, you might get help from a nearby college or university or a friendly and compassionate engineer.

■

As with sewage systems, water is either available from a water company or rural water district, or it is not. It is certainly easier and less expensive to pay a hook-up charge and monthly water bills than to drill a well, case it, build a well house, install a pump and pressure tank and run electricity.

At the same time, well water will not contain fluoride, chlorine or other additives.

A well is something we feel to be out of the category of a "do-it-yourself with junk" project. Maybe, in some places, it could still be feasible to hand dig a well, but we doubt it. Because of ever-increasing water pollution (and deple-

tion) wells need to be deeper than in pioneer days.

There are also problems with the old-fashioned cistern, and there are fewer and fewer unpolluted springs left. But water from cisterns and many springs is "soft" and wonderful for bathing and hair-washing, so it might make sense to catch rain water or pull water from a deep lake, clear river or spring, and buy bottled water to drink.

■

Water heaters are of various types, sizes and degrees of efficiency, reliability and safety.

Often, standard water heaters are removed from buildings being razed. A restaurant being torn down or gutted for another type occupancy, for instance, will produce a rather large hot water tank or, more likely, a series of fairly normal-sized ones for a residence. One never knows how much service is still in an old water heater, for sure, but since the advent of glass linings, they have been made to last longer and longer. Do be sure, however, that any tank you find didn't go through a winter, filled, in an unheated building. If it did, it probably froze and ruptured.

Electric units are easy to install because they require no venting, but they cost more to operate. Natural gas or bottle gas units must be vented properly to be safe and it is very important to make sure that a salvaged one had been used with the same type gas that you will use. Also make very sure that the automatic safety valve is on it and is functional. And, as with furnaces and boilers, a flame-fed water heater will require outside combustion air.

The older the heater you find, the less insulation it is likely to have. We don't believe that any water heaters are being manufactured with enough insulation to assure maximum efficiency. Some left-over blanket insulation can be installed around any water heater to improve it. Duct tape is not expensive and can be used for this insulating work and on a thousand or so other parts of your project.

Worth considering, too, are other ways to heat water, either directly or by use of "heat exchangers." Active solar heaters are offered by several different manufacturers at very substantial prices. In most areas of the country, the cost cannot be justified in terms of how many years it would take to amortize the installation cost. There is also, of course, the energy saving factor which different folks will regard with different degrees of importance. A conventional heating system will almost surely be needed, anyway, for sustained periods of too little sunshine.

While manufactured solar units are very expensive, building your own might cost you practically nothing. If you are a physicist, or if you have a good head for how to tinker, control water pressures, transfer heat from one place or object to another, etc., you might revolutionize present-day thinking on the subject.

A friend who lives in a mountain valley in Jamaica has an old water heater tank (stripped down to bare steel and painted black) and about a hundred feet of black ¾" pipe on the roof of his house. A pipe runs from the lowest point of the black pipe through his citrus trees and up hill to a spring behind the house next door, so pressure is kept on the system. Water in the black pipe and tank heats up and is used in the shower, tub, lavatories and kitchen sink in the house. This is the simplest system we've seen, but, unfortunately, it won't work where the temperature drops much below freezing. And not everyone has a spring above his house or lives where the sun shines almost every day of the year.

A derivative of this basic system would have an anti-freeze solution in the pipe and tank with no constant fresh water hookup. With this system the liquid in the tank would be pumped down and through a coil circling, or housed inside, an inside water tank. Heavy insulation would be essential. The heat from the coil would transfer to the water in the tank which comes from the water supply source to the bottom of the tank and goes to the hot water taps from the top.

Another take-off would be the same as above except that the pipes and tank would be on a south slope down from the house enough to allow the hot water to rise by itself to the heat exchange coil and drop when it is cooled, thus being a completely "passive" system requiring no power to operate.

In most regions of the country, heating water will sometimes require more than any solar collector could provide, because you want to take baths and wash the dishes even if the sun has been blocked by thick clouds for six days running.

So the tank with the heat exchanger coil needs to also have a burner or electric heating element below it. And if it has a burner, it *must* have a flue pipe, combustion air and an automatic safety shut-off valve to protect the lives of your loved ones!

There are other sources of heat for water. A coil of pipe built into the masonry above the fire chamber of a fireplace or in or around the flue of a wood stove or built into a wood furnace or laid in sand in a "Hahsa" unit (see chapter on "Heating, Air-Conditioning and Ventilating") will work as primary heat or in conjunction with solar-gain coils. But remember that you will need some hot water even in the summer when you won't be firing up the heating stove. Often a solar collector will furnish enough hot water by itself in warm seasons.

What we have given here is only very basic information. Some sophistication will be required for safety and efficiency, missing such items as thermometers, control valves, pop-off valves, pipe insulation, etc. But with some research at the library or university, some purely logical thought, a bunch of junk and some hard work, you can have very cheap hot water.

■

A sizeable expenditure goes for plumbing fixtures in a conventionally built house. This is especially true when quality pieces are installed.

Enameled steel tubs, lavatories and sinks don't cost much and they aren't worth much. There are situations, however, in which they will be perfectly okay — like for use by an adult or two who won't drop dishes in the sink or a glass bottle in the tub. It takes very little to knock a nickel-sized piece of enamel off and it is virtually impossible to repair the dent presentably. If you do use a "tin" fixture, be very careful installing it and finishing walls and ceiling around and over it. Many fixtures are damaged by dropped tools and materials.

Better quality tubs, sinks and lavatories are porcelainized cast iron, with some newer ones being fiberglass. Cast iron is heavy! These pieces can be damaged the same as "tin," but not nearly so easily, and "pocks" in them can be somewhat better repaired.

Many cities have repair people who specialize in filling pocks in plumbing fixtures. They are often the same people you can call to have holes repaired in leather and vinyl, so finding them in the yellow pages may take some searching. Then there are kits you can buy (usually mail-order) for making these repairs yourself, but getting the right shade of white or color won't be easy without a complete stock. At best, you cannot expect these patches to last forever. We have seen some hold for several years and others a few weeks. For this reason, it is not a good idea to buy a kit and go into business. It's one thing to catch hell from your spouse — quite another to begin getting irate phone calls at all hours of night and day.

Often the best solution to keeping damaged fixtures looking presentable is to mix some good gloss enamel to a perfect color match and keep the holes daubed full. Others feel that an antique piece with some damage shows its age and has character.

Fiberglass, on the other hand, can be repaired very satisfactorily, whether the piece is a boat, shower stall, whirlpool tub or whatever, with the proper stuff. If you find a damaged unit being thrown away or sold cheap, find the manufacturer's name stenciled on it someplace. When you find it is a Kohler, Aquaglass, American Standard, or any other brand pro-

duct, locate the distributor for that brand in your area to obtain the name of a repair person or a kit put out by the manufacturer for whatever color unit you have found. Remember that "white" can vary from the color of falling snow to yellow or grey, so the right white is essential.

There are still other shower stalls and tubs made of acrylic plastic such as those manufactured by Aquarius Industries. Some repairs can be made to these depending on the severity of the damage.

It is fun to find and reuse serviceable old plumbing fixtures. They, like other building materials, can be found at demolition sites, in wrecking company sales yards, at antique stores, auctions and garage sales.

Four-legged tubs became so popular recently that they were reintroduced in lines of newly manufactured products. They have never seemed practical from the standpoint of cleaning below them, especially if they are installed in a corner. We think they can be fun, even elegant, out in the middle of a large bathroom, but venting must be worked out. We have also seen old ones recessed into floors or into raised platforms covered with carpet, vinyl or redwood slats. Be careful about using laminate plastic (formica) for platform tops (or anyplace you might stand on it) because it can be extremely slippery.

You will probably find good old wall-hung lavatories around for the taking. These are fine for reuse as wall-hung units, but special preparations will be required to build them into vanity cabinets. In the first place, the top of a lavatory should be about thirty-one inches above the floor. If you hold your hand at that height you will swear that it is too low — absolutely! But that is where it should be unless everyone who will use it will be unusually tall or short. These wall-hung units will have front and side "skirts" of between 2" and 8" in height, meaning that the cabinet top into which you recess the unit will need to be lower by as much as the height of the skirt, so the bottom of the skirt will sit on the cabinet top. We suggest that you not try to run a cabinet top into the side of skirt or tub rim because it will not be straight, either across or up and down, making a tight fit extremely difficult to achieve, and it is virtually impossible to get a watertight seal.

There are various types of sealants to use for installation of fixtures to tops (whether using sink rims, self-rimmed pieces or old tub or lavatory units sitting on tops or platforms), but we don't think you can beat silicone.

You might find, and want to reuse, a stainless steel sink. The one you find could be practically junk, or very good, or anything in between; and the way to tell is seeing how it reacts to putting your fingers on it and how easily the fingerprints come off. Cheap units get smudges all over that won't clean off and are ugly.

Where medical offices and dental clinics are being torn out you will be likely to find some very interesting sinks and lavatory bowls. The finest sink we ever saw in a wet-bar, for instance, used to be a "cast sink" in the office of an orthopedic surgeon. It is very square-cornered, about 12″ x 17″ and has a wonderful, tall gooseneck faucet set.

Faucets can often cost as much or more than the fixtures themselves. Many people don't realize this, so will throw the faucet set out with the old fixture.

Faucet sets have been manufactured in so many different qualities that it is wise to "look the gift horse in the mouth" before going to the bother of installing one, but if the chrome will clean up and the handles turn freely, it's probably okay.

The most common problem, of course, is with the washers and the "seats" that the washers seat against. Washers are cheap and easy to install and, with a "seat wrench," so are the seats, but it is sometimes a pain to find the correct one for your set. The wrench should cost less than a dollar and will be a good tool in your collection.

Washerless faucet sets, usually with single knob or lever handles, have been around long enough now to be found in the normal "finding places." They are about the best building-

materials development since wire nails, because they are virtually maintenance free.

We witnessed a scrounger picking up a demolished tub a few months ago for ten bucks. The tub was beyond repair but it had gold-plated fittings with onyx inset handles. If coveting thy neighbor's building-materials finds is a sin, we're headed for hell!

Other plumbing fixtures and accessories to keep an eye out for are such utilitarian items as slop sinks, laundry sinks, floor drains, sprayer attachments, shower heads, hose bibs, etc.

Speaking of shower heads — we feel that a hand-held "wand," with mounting bracket, is a sensible alternative to a stationary unit. It eliminates in-the-wall plumbing and is more versatile.

And there are still other, more esoteric, fixtures you might find and want to use — such as bidets and urinals. Friends bought a house in New Mexico in the master bathroom of which had been installed a used urinal. Now that really seems dandy to me, but there is one big catch that you might find difficult to overcome: The flush valves (sloan valves) used above urinals are made to handle the water-flow of a ¾″ pipe and will not function with less. If you can run continuous ¾″ from service to fixture, you'll have it made.

Remember about pipe sizes: the designated size is the *interior* diameter. A piece of ½″ steel or plastic is about ¾″ outside, so confusion is understandable.

■

You will also want a quiet system without "air hammering," etc. Study the use of air chambers, check valves, stop valves, pressure relief valves, etc., some evening. You can find good resource material at the library and it will be time well spent in terms of peace-of-mind and labor savings. It usually takes more time to correct something than the original installation of it took.

And if you are going to install any plumbing in a concrete slab, be sure the slab is well reinforced and test all runs with *high pressure* before you pour concrete. You can rent a tester if you can't borrow one from a plumber friend.

■

Keeping hot and cold water pipes sufficiently separated will help you to have cooler water to drink.

Installation of pipe insulation on all hot water pipes will save energy and keep your water hotter, longer.

■

Common sense and a study of the principles of plumbing will pull you through your job and avoid anyone from ever being maimed by scalding water or asphyxiated by methane gas. It would also be wise to acquaint yourself with a plumbing code to learn such things as vent pipe sizes, how far it is feasible to run vents horizontally, etc.

The very most important aspect of your plumbing job will be to make sure that it is safe. This is true whether your system will be inspected or not.

■

CHAPTER TWENTY FIVE

Heating, Air-Conditioning and Ventilating

Regardless of how we all go about building our shelters, from now on it will be absolutely essential that we pay heed to energy efficiency! This is true because fuel prices are so high (and will continue to get higher), we now realize there is not an endless fossil fuel supply, and because the conversion of gaseous and solid matter to heat also expels pollutants into our breathing-air and atmosphere.

So, before studying how you will heat and cool the new house, you should figure out how to make it require the least amount of heating and cooling possible. The chapters on insulation, glass, doors, windows, and use of solar energy all relate, to some degree, to energy efficiency.

After your plan has been formulated on how to minimize loss of heat in the winter and intrusion of heat in the summer, then concentrate on ways to bring the temperature and humidity up and down as required for your comfort and health — while doing the least damage possible to your pocketbook and the ecology.

This chapter will deal with the uses of energy from the sun, earth, and water, more as means of achieving energy efficiently than as heating/cooling systems such as ones using electricity or burnable matter. We feel that the use of energy from natural sources is superior in every way and should be utilized as much as possible either for total systems or in conjunction with electricity and/or flame.

For this reason, we are devoting an entire chapter to "Use of Solar Energy."

Heating and cooling systems which operate on fossil fuels and electricity have, like automobile engines, been made much more efficient over the years following the 1973 "energy scare." In both cases, the technology was there a long time back, but most fossil fuel peddlers wanted to sell their products as fast as possible without regard to depletion, or the consumers' well-being, so that impeded advancements.

No matter where in the nation you build, how well insulated you are able to build or how much advantage you are able to take of solar or earth energy, if you borrow money for your project, you will need to include a "conventional" heating system to satisfy the lending institution. (See chapter on "Financing.")

The importance of the efficiency of this system depends on how well your house will do without it. If it will be required to run half of the time throughout the winter, you had better buy a new system. If you have your plans reviewed and a heat-loss computation run by a mechanical engineer who tells you that the back-up system will only kick on three or four times a year, you might find a used, less efficient furnace somewhere. (But see what the money-man and codes have to say first.)

We are less and less thrilled with air-conditioning as fuel costs get higher and higher. There are what we consider to be viable alternatives to air-conditioning which we will discuss later in the chapter. Man used fire to warm his cave a very long time ago while cooling his surrounding air became commonplace only since WWII. The point being that we must have heating, while cooling is just "nice."

In most areas of the country, electricity is a derivative of fossil fuel — so we are going to include it when we use the term "fossil fuel."

With the use of electricity for thermostats and fans, we can use fossil-fuel heating systems safely and automatically.

Home ownership is a beautiful thing, indeed, and one can often derive extreme satisfaction from building with one's own hands. However, it is easy to fall into a trap wherein you are sometimes a slave to your house!

Heating with wood (more on this later), coal or with kerosene heaters has some people tied down to their property like milk-cows have others. It can be a rewarding experience to keep the furnace stoked with fuel. It can be anything but rewarding to have to miss a trip to Miami because if you're not home, the fire will go out and the pipes will freeze — or to have to hike five miles through snow drifts to recharge the heater instead of spending the night in town during a blizzard.

Electric heating is available in all sorts of systems — all of which are expensive to operate. They can be thermostatically controlled to keep the house at 72° — or to kick on in emergencies to protect against freeze. Electric heating coils come in about any imaginable size for installation in walls, baseboard areas, floors and ceilings, and there are portable units.

Heat from electricity can be "radiant" (without fans) or with fans spreading the heat outward from the source. Either way, it is clean, comfortable and, if properly installed, safe.

Fuel oil and bottled gas (propane and butane) are also expensive but not generally as high per BTU as electricity.

Oil can be dirtier and, like gas, requires proper venting to be safe.

In remote locations and areas of the country where electricity is unusually high and natural gas is not available, using oil or bottled gas makes sense for back-up or auxiliary heat.

Your neighborhood oil or gas company will help you figure out what you need and furnish you a tank or bottle.

You will still need electricity to run the fan and thermostat. Keep in mind that a power failure will (to assure safety) automatically stop the flow of fuel to the burner. Please, under any circumstance, do not install any kind of gas or liquid burning device without a tested, functional, automatic cut-off valve.

Also, take care to provide combustion air from outside to the area of the furnace, boiler or water-heater and install a safe flue! If you are not completely knowledgeable about burners and flues, it is imperative that you seek the help of someone who, for certain, is.

Another warning — be sure you don't install a salvaged heating system which was manufactured to use a fuel other than what you will be using, without proper modifications.

Natural gas, where available, is much less expensive than electricity, oil or bottled gas, at the time of writing. While natural gas is less expensive per unit, this price difference is partially offset by the fact that it does not burn as hot as bottled gas — thus requiring more to do the same job.

As with oil, bottled-gas-burning furnaces, boilers and water heaters, natural gas units must have safety cut-off valves, combustion air vents, proper burners, and safe flues.

Heat-pump heating and cooling units are now in fairly common use and are more efficient than conventional equipment, thus less expensive to operate. Heat-pump furnaces will not work where temperatures are too low, however. You will not be too likely to find salvaged heat-pumps yet.

In many areas of the country, a residence constructed with good energy conservation methods will not need a regular "forced air" heating system to prevent water freezing and to provide a family sufficient heat for health and comfort.

Portable electric heaters, gas circulating stoves, oil heaters, wood and coal stoves all have advantages for certain uses in the various areas of the country. They will probably not qualify as "heating systems" for purposes of financing and, again, there is no way to over-emphasize the importance of proper install-ation and venting. These free-standing units are available everywhere — in garage sales, auctions, second-hand stores, etc.

Wood as fuel has one tremendous advantage over fossil fuels in that the supply is re-plenishable.

If you are building out in the sticks where you can harvest your fuel from your own land, you will heat for the cost of chain-saw fuel, re-pair costs and hard labor. On the other hand, if you heat with wood in a metropolitan area, you might have to pay a price for wood ap-proximating that of natural gas without the conveniences of automatic feed and thermo-stat controls.

Fireplaces without inserts are pretty to look at and nice for roasting chestnuts, but they are not really satisfactory for keeping a house com-fortable. A fireplace with an open front and no source of exterior combustion air will, when being used, burn the warmed (tempered) air from the house and, while not in use, allow warm air to be drawn up the flue and lost. If you do build a masonry fireplace, look into in-stalling an insert (with a pipe for bringing outside air in for combustion) and a plenum and blower-fan system. (Not very pretty, but practical.)

"Tin" fireplaces (little free-standing things) are available, used, in abundance because folks have replaced them with real stoves. They have the same disadvantages as masonry units, except more so.

Wood stoves are big business nowadays and are manufactured out of everything from sheet-metal to heavy cast-iron with firebrick linings. Generally speaking, you get what you pay for and the heavier the unit the better the quality in terms of both durability (light metal will burn out and rust) and efficiency (weight collects heat and dissipates it more effectively and evenly).

There are more and more used wood stoves for sale all the time, for two reasons. First, some people who installed them found that their ambitions or life styles were not condu-cive to keeping them in use. Second, some people have had what they perceived to be close-calls with burning their houses down be-fore they learned to regulate their rate-of-burn or they are scared by the fact that so many fires are caused by wood stoves.

It is absolutely true that one needs to learn how to operate a stove properly and maintain it. Periodic cleaning of the flue is vitally impor-tant as is proper flue installation. Closing the stove down too much, thus burning the wood too "cool," causes creosote to deposit in the flue and sometimes run out at the joints. This cre-ates an extreme fire hazard. One very impor-tant rule to follow in flue installations is to make all joints with the pipe on top fitting down into the section below so the "tar" cannot run down and out.

Wood burning furnaces are being factory manufactured and hand-made all across the country. Some even have thermostatically con-trolled dampers, etc., and mechanical feeders.

We have seen some real Rube-Goldbergs that are functioning very well and apparently safe. We have seen others that scare us to death.

If you are inclined to build a wood furnace, be on the lookout for fire bricks (cream-colored-solid bricks, usually with the manufac-turer's name embossed), steel angles, channels, etc., and anything you can use for stove "hard-ware." We have seen units built out of old boiler tanks, fifty-five gallon steel drums, and old water-heater tanks out of laundromats.

Common sense tells you that any furnace must be installed and vented in such a way that it cannot overheat any combustible part of the structure.

Your library will have books dealing with design, building, installation and operation of wood furnaces.

For those of you who are such "purists" that you don't even want to use electricity for air movement, you might consider building your living quarters above your furnace and allow the heated, lighter air to rise through ducts or grates, as was done so much around the turn of the century.

There is one other type of heating unit which we feel to be worth mention. This is a remote-from-the-house building which acts as an incinerator and is connected to the heating system of the house by a buried, insulated heat supply pipe. An outfit by the name of HAHSA at P.O. Box 112, Falls, PA 18615, sells these things in kit form and some people have studied the literature and built their own.

A big advantage that we see to this is that anything which burns (trash, newspapers, etc.) can be fed into it to produce heat and save trash collection expense. It does, however, require periodic cleaning-out.

It is built in such a way that it will pass all codes we are familiar with, as an incinerator.

There are types of heating systems other than those mentioned above. We have seen hand-made systems which work with hot and chilled water, steam, well-water, etc. Old radiators are plentiful in junk yards. If you have interest and ability in working with applying basic laws of physics to pipes, valves, heat chambers, heat exchangers, etc., have at it! We wish you great success and ask that you send us pictures and data sheets.

Or you might decide on an alternative means to keep reasonably comfortable in the summer.

Conventional air conditioners are, essentially, dehumidifiers. In locations of very low humidity, the most common alternative, the evaporative cooler or "swamp cooler," makes good sense. This is simply a device in which a fan blows outside air across a pad saturated with moving water and into the building. The movement of moist air is comfortable if the moisture content of the normal air is low.

Increasing humidity when it is already high does little or no good for comfort and can cause mildew to form on clothes, etc.

Another technique of cooling which is catching on is a way of using the constant, cool temperature of the earth. By burying a pipe of some sort in your yard and moving air through it into your house you can produce a movement of air which has been cooled.

The air can be routed into a duct system (the same one your furnace uses is okay) or through one central air grille or register.

Cooled air, unlike hot air, is "lazy" and will tend to settle. For this reason more velocity is required to move cool air sufficiently to produce comfort. As a rule of thumb, the higher the CFM (cubic-foot-per-minute) of cool air movement the higher degree of comfort. This rule tends to break down a little when you can't keep your hat on in a gale.

CFM will, of course, depend on fan size. To give you some idea, an exhaust fan in the ceiling of a bathroom is about 180 CFM, while the fan supplying air-conditioned air to a three bedroom home might be around 1000 CFM.

In a buried-tube cooling system, the amount of cooling will depend, in large degree, on the diameter and length of the pipe.

There are certain fundamental rules to follow in the installation of this type system. As with any conduit of moving air, it is good to avoid changes of direction in the pipe as much as practical and it is necessary to cover the outside end of the pipe with protective grates and screens to keep the neighborhood cats, dogs, mice, birds, and spiders from seeking cool haven. Also build-in a removable air filter near the house end if you don't use the furnace filters.

It will pay to find where the buried utilities are in your yard so you can avoid digging into any. And get good connections between pieces of pipe to keep water and worms out. Steel stripping, cut-up innertubes, and roofing mastic are good.

Culvert "whistle" pipes are often removed in reasonably good shape, and disposed of, as are various other kinds of big ducts, pipes and tubes. We recently obtained, free, about twenty-five feet of new green fiberglass, nineteen inch diameter pipe left over from an under-slab return air system at a new church construction site. And steel drums, if properly jointed, make good tubes.

The crawl-space below your house, if you build that way, can furnish a supply of cool air. On one of our recent projects, floor openings were installed to the inside of a greenhouse room floor. These openings turn and go through holes cut in the rim joists of the house, into the crawl space. In this case, an old attic fan was mounted at the outside access door to the crawl space. When the fan is turned on, and the foundation vents are closed, a "positive pressure" is built up in the crawl space and gobs of cool air pour into the greenhouse and on into the house through three patio doors. Provisions had to be made to keep heat from building.up in the greenhouse — more about that in the section below dealing with ventilation and in the chapter on "Solar Energy Use."

As mentioned earlier, in the chapters on "Plastics" and "Obtaining and Storing Materials," it is desirable to install polyethylene plastic sheeting over the earth below your house to block excessive moisture from coming in. If you create air movement over this sheeting, it will need to be weighted down to keep it in place.

A fan could have been installed, at the opening in the floor, which would pull the air through the crawl-space with the "make-up" air coming in through open foundation vents, but a scrounged attic fan was on-hand.

Keep in mind the importance of providing some means of opening and closing holes you build in around your structure. You don't want to install a device which provides comfort in one season and is a nightmare in another. We think that a piece of insulating board in a hole is the best means of closing, if it is fairly accessible. Building demolition sites and junk yards often yield grilles, grates, shutters, and access doors.

Insect screening of holes is also important and, in some instances, air filters can be beneficial.

When a really good supply of cold water is available from a lake, spring, pond, river, or large well, it can be utilized for cooling.

If you have a good grasp of physics and plumbing, there's no telling what you might come up with using junk, scraps and salvaged parts. Or you might want to hire a mechanical engineer to come up with something for you. Except for possibly flooding yourself, or building up too much humidity, you won't face the hazards with cooling that you will with heating systems — so you can be more adventurous.

■

Ventilation is an extremely important matter! Provide proper ventilation in your well-insulated house and you will be protected from the bad effects of too much moisture and overwhelming summer heat, even without air-conditioning.

All you need to keep in mind in order to design yourself a good ventilation system are a few very basic facts:

1. It is difficult to get too much.
2. Hot air rises.
3. Air movement feels good when you're trying to keep cool.
4. When air is forced into an area, there needs to be some way for it to escape.
5. When air is forced out of an area, there needs to be some way for it to be replenished.

With these things in mind, you can figure out where heat will accumulate and devise a system for dispelling it.

Attic space is the most important area we know of to keep properly ventilated. Tremendous heat can accumulate there and permeate downward into the living space.

The most common ventilating device for cooling is the attic fan, and we highly recommend its use. New attic fans are not expensive and used ones are often available.

Units manufactured for conventional attic installations can be used in many other ways.

One project we know of is a geodesic dome with a cupola built on top. An attic fan is mounted in an enclosure which hangs below the louvered cupola. An insulating shutter built of styrofoam panels opens and closes automatically below the fan.

We have seen "attic fans" installed in walls and crawl spaces as well.

There are many types of attic vents on the market, including gable-end louvers and roof vents, but the best, by far, in our opinion, is the continuous-ridge-vent used in conjunction with vents in the soffits around the house. Be sure that ceiling insulation is not blown or installed over the soffit vents.

When an attic fan is used, it is important that enough opening is provided to allow the air you are putting up there to escape. If gable-end louvers are installed, it might be good to cut some pieces of plywood or something to cover them from the inside for winter.

Crawl spaces should be provided with vents which can be opened for summer and closed for winter.

Exhaust ventilators in bathrooms and over kitchen stoves are fine except that they pull warm air out of the house in the winter. You might decide to put up with the odors to conserve energy.

Fans are available in all sizes and shapes, in propeller type or squirrel-cage, new and used. Some operate on 110 volt electricity, some on 208 or 230 single-phase and some on 208 or 230 three-phase. Three-phase motors are more efficient to run, but it is not likely that three-phase current will be available to you.

Power roof vents and wall vents are being removed from old buildings and you might get them free. Gravity type (non-powered) exhaust vents are also being discarded.

When some of this type of equipment falls into your hands for free, or nearly so, take it home, study it and find a use for it or sell it to someone who can. Or trade it for something you need more. The very least you can get is scrap metal price at the junk yard. But don't get rid of a single-phase blower until the tail end of your building project, because you might see ways to move air from one space to another to produce comfort.

CHAPTER TWENTY SIX

Uses of Solar Energy

Because this is not a technical "How-to" book, this chapter has been included only to stimulate the reader's interest, offer very basic guidelines, and encourage everyone to use, as much as possible, his or her share of the enormous and free energy the sun gives us.

We hear the terms "passive" and "active" applied to solar heating systems and these are what will be discussed here.

The vastness of the North American continent makes for extreme variation in temperatures, hours-per-day of sunlight, sun angles, etc. These variations mean that passive systems are more feasible in some areas than others, for instance, and the high cost of an active system can be easier justified in an area where cold is extreme, and sunny days are the norm, than in more mild and cloudy areas.

■

The idea of a passive system, simply put, is to allow sun rays to enter a space, land on a surface which is dark enough in color to absorb the rays, allow heat to accumulate in the matter of the surface (mass) and dissipate from the mass to warm the space.

Since there is no heat involved with sun rays until they are absorbed by something, color is of extreme importance. Black will absorb the most while a white surface will reflect about 90 percent away. If sun shines through your window, reflects off a white wall and back out again, it has done practically nothing to raise the temperature of the room.

And if sun rays come in and hit dark paint, most of the heat from them is "caught" and immediately dissipated into the air because there is no substance to store it.

On the other hand, if the sun rays come in and land on a foot-thick masonry wall, with a dark exposed surface, that surface will absorb most of the heat, which will penetrate into (perhaps through) the masonry. The advantage, obviously, is that this absorbed heat will flow out of the mass and into the enclosed air of the house long after the sun is no longer shining in.

However, as we mentioned in the chapter on "Glass and Mirrors," glass, even double- and triple-pane "insulating" glass, has precious little insulative value. So the house with the heat pouring off the sun-heated, interior masonry wall, will have heat pouring out to the great outdoors, through the glass area which allowed it to come in — *unless* something is closed in front of the glass which will impede the outward flow.

There are some new fabrics which have very good insulating capacity for their thickness, but this is thin stuff and, used as drapes, does not provide much of a sealed barrier.

We prefer shutters or panels which can be closed rather tightly over the glass area. The chapter on "Insulation and Vapor Barriers" mentions several products which could be used to build these panels — many of which you just might find while scrounging.

Possible hardware for slide panels, hinge shutters, etc., has also been discussed.

Manually opening and closing insulation barriers is what is usually required, but there is the obvious drawback of someone always being needed at home to monitor the sun and operate the system.

A beautiful alternative is the use of an electrically operated opening/closing device controlled by a photo-cell. Purists might belittle this departure from a purely passive system, but the cost of the tiny amount of electricity used to power the mechanism can be saved many times over — not to mention the convenience.

For some reason photoelectric cells, to monitor the intensity of the sun, are almost always available through one kind of a "deal" or another. Electrical equipment companies, surplus equipment companies, photographic supply houses, electric utilities companies and farm suppliers are places to contact.

Fractional horsepower electric motors are not expensive and can often be scrounged from junked furnaces, fans or a myriad of other often-discarded items. So are switching devices.

The hardware for rolling, hinging, sliding, or whatever you design for your particular need, can be out-of-sight expensive if someone is hired to engineer and build the system. If scrounged, these materials don't need to cost much, if anything. The trick is to design a system in general terms and then go out to junk yards, sales and trash heaps to find items which can be incorporated into your general design concept.

There are hundreds of books available at bookstores, libraries and alternative-energy-equipment stores to use as guidelines for rock-beds, duct systems, glazing, shuttering, "Trombe-Walls" and all the rest of the passive solar components thought of to date.

But remember this — man's recent resurgence of interest in the use of the sun's warmth to keep his dwellings comfortable has given birth to a new applied science which is still very much in its infancy.

Any infant makes many mistakes and has a lot to learn. So when you are studying the passive-solar-house books, take a hard look at the *results* achieved by the different approaches. And consider how you can improve on someone else's ideas. *Then*, figure out how you can create your own system, better than anything in the books, out of what you have been able to scrounge! Let others study your system and improve on it. So goes development.

■

Active systems require "hardware" to do two things. Something is needed to collect solar heat and something is needed to move it from where it is collected to where it is to be released in the building.

Air collectors (such as flat-plate collectors) and fluid collectors are available from several different sources, or can be handmade — often from scrounged materials.

Often, collectors can be arranged to collect heat at a lower elevation than the part of the house to which it will be piped or ducted. If allowed to travel by gravity-flow, it is technically a "passive" system. Some flat-plate collector systems used this way are so incredibly simple and effective that it is a wonder we weren't using them even when fossil fuels were cheap.

Active air systems usually require fans, ducts of several sizes and shapes, dampers, thermostats, etc.

Active fluid systems will require pumps, coils, tanks, control valves, check-valves, safety valves, pipes, etc.

It is amazing how much good stuff is being thrown away — high wages render it unfeasible to salvage. We were recently given eleven roof-top air-conditioning units, each of which contains two copper-coil radiators, several valves of different types, pipes, fittings, motors, fan blades and a compressor.

The compressors are probably all shot and the motors are generally unusable, three-phase type, but the rest of the stuff can be incorporated into heat exchangers, energy moving and controlling devices, etc.

As with passive system design and construction, much information is available on active systems, with even more room for advancement and improvement.

Systems for heating domestic hot water have been somewhat more refined and are also considerably easier to build than space-heating systems. These are discussed in more detail in the chapter on "Plumbing."

There are many other possible uses of solar energy to create comfort and reduce expense — air-conditioning, producing electricity with photovoltaic cells, indoor gardening, even food drying — and we heartily endorse the incorporation of any of them (and anything else you can think of) to beat the system of paying astronomical prices for the privilege of helping to deplete the world's fossil-fuel supply.

It cannot be over-emphasized, however, that the greatest system-beating feature you can incorporate, in this regard, is that of making your house as well insulated and sealed as the climate in your part of the world dictates.

While it is extremely satisfying to have created a residence for yourself, at a fraction of normal cost, it is doubly gratifying to have created it in such a way that it can be kept comfortable for your family at a small fraction of normal utility costs!

Closable, high "R" value, insulating panels in a house cantilevered into the trees.

■

CHAPTER TWENTY SEVEN

Tools and Equipment

Even the world's finest craftsmen and mechanics need good equipment to ply their various trades. They won't always agree on exactly what tool should be used for a specific task, or even how it should be adjusted, or held, but there will be universal agreement that, whatever the tools, they should be maintained in perfect working order.

Later in this chapter, we will discuss types of tools and equipment pertaining to the various trades you might be performing on your project. But, first, a few words on acquiring some.

For some strange reason most men love to buy tools at auctions! Consequently they usually command a rather high price — sometimes ridiculously high. This is particularly true of any handtool with even a hint of "antique" about it. But occasionally some tools will show up at a sale devoted mostly to other types of stuff and sell very reasonably — possibly even cheaply. The best protection against being caught up in buying fever, and wasting money, is to know current, new costs.

Buying electric tools (drills, routers, sanders, etc.) at auctions is a risky practice, even when you are allowed to plug them in and run them — unless you can put pressure on them in a fairly quiet area to feel lugging-down, hear loose bearings and see sparks flying. This is not likely to be possible at an auction site.

It is really true that "they don't make them like they used to." It is also true that they don't make parts for them like they used to — and the very best tools will eventually need repair.

An old contractor-friend died several years ago and his widow had his workshop contents auctioned. Some wonderful equipment, in tip-top shape, passed on to some proud, new owners. But nobody knew that old Charlie couldn't bear to throw away any old tool. When a piece had been repaired several times, was getting sluggish or noisy again, and was replaced, it was put aside for an eventual emergency use. A lot of obsolete, all metal, ball-bearing, brand-name pieces became the property of people who were not so very proud.

We have known a few tool-fanatics who, with their collections, were "tool-poor," but they are a very rare breed. We have, on the other hand, known hundreds, including ourselves, who have struggled pathetically with a task, because the right tool was not at hand.

It would not be possible for you to list, or even imagine, every device you will use on your project (or will, if it is available for use). If it is in your tool pile, you will use it — from a six-foot railroad wedge-bar to a dentist's cavity pick — we guarantee it! For this reason, we encourage scrounging all tool-type things which come your way free, or nearly so.

Now, a few words about tools that cut materials (saws of all kinds, drill bits, knives, shears, etc., etc.). There is very little more productive time spent around construction than that used for sharpening cutting edges. This is true for several reasons, the most important one being safety.

Generally speaking, sharp tools do not cut people. Dull tools, being forced, slip, break or bind and cut the hapless users.

The writer knows from many years experience in a woodworking shop that dull chisels and yankee (ratchet) screwdrivers cause more injuries than all the power equipment combined.

The speed at which a tool cuts is proportionate to sharpness, so most time spent sharpening, or taking tools to sharpeners, will be made up.

Sharp instruments also conserve the power-source, whether that be your body or an electric motor.

And the finished product will certainly reflect the sharpness of tools used to produce it. Dull saws (hand or electric-powered) will not cut straight lines because they will "lead." So will drill-bits, and they will leave ragged hole edges.

■

Most new-construction projects almost demand the use of one fairly expensive tool to allow "getting off on the right foot," and staying there, in the early stages. A farm level, with a tripod and measuring rod, will let you establish footing depths, determine elevations of foundation walls, drives, drainage pipes, etc., to insure level floors, properly functioning septic-tank lateral fields and sewer lines, ground waters flowing away from the structure (instead of into it), and gobs of other less dramatic things.

You won't need a transit (which will be okay if one is available) and you will find a simple level easier to set up and less intimidating. A new level will come with easy-to-follow instructions on how to set up and use it. If you can scrounge an old one (and they are available at greatly reduced prices), the library will have instructions. We know that words like "transit-level" and "surveying" sound overwhelming to some people. The truth is that you can learn what you need to know to use a level correctly in less than an hour from a book or friend.

■

General-duty tools and equipment — wheel barrows, ice pick, magnet, picks, shovel, ladders, saw-horses (build your own with salvaged off-falls), plan table (do the same), flashlight, water hose, sledge hammer, pocket knife, assorted screwdrivers, garden tools, clamps, pry-bars, wrenches (including a crescent or two), pliers, brooms (straight and brush), and hatchet will be used during virtually all phases and by all trades. It would be wonderful (and probably impossible) if all these items could be kept, unbroken and unlost, in a safe place every evening after work ends. Good luck!

■

A very special category of hand tools is worthy of being emphasized here because of the many hours of extra labor, the frustration and the goof-ups that will be avoided if liberal use is made of them. These are the instruments which, along with the aforementioned farm level, will help to keep the project square, plumb and level.

A few types of structures, such as geodesic domes and yurts, are erected with level floors but irregular wall/roof planes. Because of the irregularity these are nightmares to build interior walls to and trim-out.

When something will go together easily, although a little out of square or leaning, it is dangerously easy to tell oneself that "I'll take care of that with a trim piece." Believe me, fitting the "trim piece" will take twice the time and effort as the initial correction would have, not to mention the cost and possible less-than-satisfactory appearance.

A good bar-level, at least 24″ long (masons use twice that length), can be used to erect materials both plumb and level. Examine used ones closely before buying, to see that the air-bubbles are no longer than the closest-together truing lines and the glass lenses aren't broken out.

Framing squares can get bent. Find a good true one and protect it. You will be amazed how often you will use it — not just for framing.

Speed squares are great for guiding electric rotary saws (Skil saws), as well as many other jobs.

The study of all the possible uses for framing and speed squares is a fascinating one but probably more time consuming than worthwhile except to learn lay-out of rafters and stair carriages.

Combination squares are good to use for smaller work and for 45° cuts. They are also great for using as marking and offset gauges.

Plumb bobs are sometimes useful but not nearly as necessary as levels and squares.

■

Measuring devices have become much more sophisticated and easy to use over the last ten to fifteen years. Measuring tapes, small enough to hang on your belt, are now readily available up to thirty feet long. They are ¾″ wide and have enough rigidity to allow one-person measuring of wide and fairly high areas.

Usually marked with red or black at each sixteen-inch increment, in addition to the feet and inches, they make measuring of standardized materials even more convenient.

Some tapes also have conversion tables, nail lengths, etc., on the reverse face — sometimes extremely helpful.

If you have, in addition to a tape, a six-foot wood folding rule with a slide-extension, you will find many uses for it, such as measuring depths. These are often available at auctions and garage sales.

■

Concrete work requires very little equipment besides a cement-mixer, and that only if you do not buy from a ready-mix company. Watch out for some trowels and an edging tool and sort out a few straight 2 x 4′s, of various lengths, to use for screeding.

If you have obtained a cement mixer, it can also be used to mix mortar for any masonry work you might do. Or you could simply rig up a mortar trough and mix with a hoe.

Again, watch garage and junk sales for brick trowels, wire brushes, joint raking tools and masonry hammers.

■

Standard carpentry — construction of run-of-the-mill houses with standardized materials — is a learned science. Carpentry applied to personalized, interesting houses is an art which requires combining the uses of logic, innovation and creativity.

Innovative carpentry calls for the use of several standard-type tools and, possibly, some improvised ones. In addition to the "general-purpose" tools mentioned earlier we recommend these:

Hammers — one twenty-ounce, rough-faced for framing and one sixteen-ounce, smooth-and-convex-faced for trim. Steel handles are good for the nail-pulling and prying you will almost certainly be doing.

Hand saws — An eleven-point for trim-work and an eight-point for rough stuff.

Hand planes — a "jack-plane" about 10″ long and a "block-plane."

Wood chisels — the more you can latch onto, the better.

Hack saw

Tin snips

Stapler which will shoot 9/16″ staples

Nail/tool belt

Vise-grip pliers

Channel lock pliers

Nail puller

Ratchet (Yankee) screwdriver if you won't be using an electric drill motor. Use with extreme care.

Nail sets — one for small nails and one for large.

Utility knife

Wrecking bar — about 24 to 30 inches long.

Brace with several bits (if you will not be using electric power-tools).

Bevel square (sliding "T" level)

Chalk-line box

Putty knives — a narrow one (about 1″ wide) and one about 4″ wide.

This list is, of course, quite general and not in any order of importance. There are items included which you could get by without — particularly for certain types of structures. There are other items, such as hand-miter box-saws, which are wonderfully convenient if the pocketbook can stand it or if one can be scrounged.

■

Power tools can certainly save untold hours of labor and gallons of elbow grease. We recognize that some projects are in the sticks where there is no electricity (although there are rechargeable battery-run tools available now) and there are some people who want the satisfaction of having done everything "by hand." We are convinced that the use of power tools does little to diminish the "by-hand" label — there will definitely be enough hand work for the project to quality for it.

Circular saws (Skil saws) and drills are the two most-used power tools. Carbide tipped blades are wonderful for the saw if they can be kept out of steel embedded in the wood being cut. And there are all sorts of special types of blades for concrete, plastic, metals, etc.

Good ⅜″, variable-speed drills with reverse action are available at fairly reasonable prices during promotional sales and, sometimes, at auctions and garage sales.

All types of bits are available to drill wood, metal, concrete, and plastics. Screwdriver bits save hours of work. Specialty items such as grinder wheels, buffing pads, sander cylinders, wire brushes and paint stirring arms are often

to be found in miscellaneous-junk-boxes at sales and can be very handy if you can remember where you put them.

A saber saw (Jigger saw) can be another tremendously handy tool around any building project and especially if there will be any curved detailing to cut.

Routers, too, save lots of time and effort and an unimaginable variety of cutters are available to do anything from placing tongues-and-grooves on board edges to edging formica.

Belt sanders make for professional-looking projects, but they require some learning-about and practice before they can be very helpful. We suggest practicing sanding blank boards before applying a belt to a piece of millwork.

Oscillating and vibrator sanders save time and hand muscles and help produce extremely well finished products. They are slow and not suited for sanding-out irregularities but are superior to belt sanders for final preparation of wood to receive transparent finish.

The rest of the hand-held electric power tools all have merit for specific types of jobs, but can probably be just as well forgotten about for a single, custom project.

Of course, if your uncle has something he doesn't use or if a sale turns up a hand power-plane, heat-gun, grinder/polisher, mini-rotary-tool or reciprocating saw, don't let it get away. Whatever it is, you will use it at least once and wonder what you would have done without it.

There are stands, adapters and accessories for all hand power tools. If you have them, fine. If not, look at a Sears catalog for ideas and figure out how to create your own, special tool for your own special job.

Bench grinders are wonderful things for many purposes — including tool sharpening. There are millions of them lying around unused.

Air power tools are also available (and often much cheaper than electric ones) at auctions and special sales. The drawback is that they require a compressor with hoses, adapters, pressure regulators, etc. If you are fortunate enough to have access to a compressor, look into air-nailers, staplers, saws, screwdrivers, drills and sprayers.

■

Floor-standing power equipment can be wonderfully handy, but is generally expensive. Aside from a bench grinder and possibly a radial-arm (cut-off) saw, it would probably not be worth the expense and bother except for building cabinets and furniture.

A table saw (with some carbide-tip blades), a jointer and a drill-press are about all that are essential for a mechanized cabinet shop. The saw, however, must be of sufficient size (10″ blade capacity) and with enough power (a minimum of 1½ H.P.) to be a real piece of equipment.

Jointers are sometimes available at sales, etc., without guards. Please don't even start to use a jointer without a spring-loaded cover-guard and even then, be careful. Jointers are involved in more serious accidents than any other of the woodworking power tools.

Band saws and jig saws are needed for some types of projects, but are less important than table saws for most. Bench sanders with belts and/or discs, shapers and even planers can be great additions to the basic three-piece shop if they come your way.

We consider a lathe to be a highly specialized piece of equipment not worthy of the space it would occupy, unless, of course, your project will require lots of turnings.

■

Roofing, depending on the material used, is most easily accomplished by using some specialized (usually very modest) tools. We think that watching experienced craftsmen and studying their tools and movements is time very well spent.

Insulation and vapor-barrier installation, depending on what kinds of insulations have been scrounged, requires little more equipment than a utility knife, a pair of scissors, or tin-snips and a hand stapler.

■

Metal working varies so much with types and thicknesses of stock that a list of required tools would be meaningless. But for most people using heavy steel (beams, columns, etc.), the best bet is to have it cut in a fabrication plant where there are acetylene torches, power-band-saws and the like.

Depending on the quantity of holes to drill, it might be advisable to pay for shop-drilling, but a good drill motor and quality bits will let you do a lot of your own drilling.

Portable welding machines can be rented, in some areas, for those who know how to weld and torch. There are also people who will make "house-calls" with portable rigs. Our experience has shown that the one-man operations often charge reasonably, due to minimal overhead and the fact that these are individualists who tend to go easy on fellow, struggling individualists.

Amazing things can be accomplished with light metals and ordinary work-bench tools such as a hacksaw, files, sabre-saw with metal blades, tin-snips and a drill.

Soldering irons can be fun to use if you're into that kind of thing.

■

Glass cutting is a skill well worth learning. Again, some time spent watching people in the workroom of a glass company will be an entertaining, informative experience. Build yourself a carpeted table, buy a new glasscutter, and practice on scraps before starting the real job.

Plastics require sharp tools. Saws should have many-tips-per-inch. Cutting plastics, more than any other phase of the project, requires wearing goggles. Feed tools slowly and steadily while holding stock as firmly as possible at the point of machining.

■

Watching the techniques of painters and paper-hangers and, if possible, working with them a few hours at the various types of jobs will allow you to develop your own ways to do your work and pick out the tools most suited to you.

While scrounging, watch for good paint brushes, wall-paper brushes, roller handles, etc., and accumulate film-plastic and sheets to use for drop cloths.

If your equipment includes an air compressor, it should also include a spray-gun. Spraying paint, stains, clear finishes, etc., is *so much* faster and more even.

An airless-sprayer is also a great tool if it is large enough to be more than a toy. A $1/3$ horsepower unit with five quart container is very satisfactory for residential work.

■

Flooring materials are so varied that tools and equipment needed for installing many types have already been mentioned.

Resilient materials (tiles and vinyl "linoleum") are laid on a bed of mastic which is applied with a notched trowel. Cutting can be done with a knife or heavy scissors.

Carpet laying (either glue-down or with tack strips) requires considerable stretching to achieve a satisfactory job. It is possible to concoct a carpet stretcher, but it would probably be worth the dollars it would cost to rent one at a rental store or from a carpet company. If carpet is not well stretched, it will expand and hump-up in ridges.

Carpet knives are available, and cheap, where you buy supplies like double-face tape, seaming tape and tack strip.

Ceramic and quarry tile work requires equipment which can be rented but is really too expensive to buy new and is not likely to be found being discarded.

■

Electrical wiring, aside from conduit work, can be done with a pocket knife, screwdriver and pliers. A few tools for specific purposes, however, will drastically cut the number of man-hours and frustrations.

The best electrician's tool investment we are aware of is a combination cutter/stripper/crimper. This may not be a scroungable type of item and, at the time of writing, it costs about ten bucks. Well worth it!

There are also items you can scrounge for your tool belt. Side cutters, needle-nose pliers, various other pliers and a wide variety of screwdrivers will speed your project along.

Not a tool, but a timesaver and wonderful product, is the wire-nut. Manufactured in lots of different sizes and fairly inexpensive (also easily scrounged from demolition sites) they are good to keep, in various sizes, in the tool belt. They are better than tape for wire joints anyway.

Using conduit requires cutting and bending. Your general-duty hack saw will cut the conduit, but a conduit bender will also be needed. This is a very simple device, with no moving parts, which commands a ridiculous price when purchased new. Keep an eye peeled for one being discarded.

Finally, a simple plug-in tester, very inexpensive, will help to determine if your job is in perfect working order and, if not, what wires were crossed or left loose.

■

Plumbing tools vary greatly with the type (or types) of materials going into a project.

Plastic piping requires a hack saw and a wrench or two if threaded fittings are used.

Steel pipe, on the other hand, must be cut, threaded and put together with fittings.

Copper requires cutting and sweating — so a propane torch with accessories is needed.

Pipe wrenches, threaders, chain vises, pipe and tube cutters, seat wrench, basin wrench, slip and locknut wrench, flaring tools, etc., are available at auctions and close-out sales.

Pipe joint compound, solder and plastic pipe cement might also show up in miscellaneous junk boxes.

■

Heating, ventilating, air-conditioning and solar energy equipment installation require many of the same tools used in other types of work mentioned above.

Metal duct work, however, can be made easier and better with marking devices, crimpers, tinning hammers, etc.

■

There are some good old standard brands of tools and equipment that are definitely superior in quality. Many of the new, cheap tools are trash which can bring more frustration than help.

In picking up old tools, look for names like "Disston," "Buck," "Blue-grass," "Stanley," "Crescent," "Yankee," "Marples," and "Sheffield Steel."

"Craftsman" on hand tools and "Skil," "Black and Decker," "Milwaukee," "Craftsman" and "Stanley," on portable power tools denote quality.

Floor-standing equipment bearing names like "Rockwell," "Wadkins," "Unisaw," "Delta" and "DeWalt" will be good pieces if they have been reasonably well maintained.

The same names on new stuff mean quality, generally, but there is another element to consider — "Craftsman" tools sold by Sears-Roebuck have great guarantees, lifetime on some items. And, so far at least, Sears has stood by their guarantee to the point of no-questions-asked. We are not plugging these tools because they are better than the others mentioned above, but because they are consistently good quality, fairly priced, available everywhere and guaranteed. This cannot be said for many other lines of tools.

Wherever you buy new tools, particularly power tools, watch for sales — they sometimes offer substantial savings.

Index

Abitibi board, 70
Acoustical ceiling panels, 78
Acquiring materials, 5-12
Acrylics, 90
Adobe, 33
Air collectors, 146
Air compressor, 154
Air conditioning, 142-143
    cold water, 143
    from buried pipe, 142-143
Air filters, 143
Airport terminal seating, 119
Aluminum pie plates, 84
Antique cabinets, 103
Antiques, 118
Appliances, 123-126
    efficiency, 123
    safety, 123
    wood burning, 125
Architect, 20
Aromatic red cedar, 47
Art, 122
Asbestos, 11
Asphalt roofs, 75
    shingles, 72
Attic fans, 143-144
Attics, 80
Auctions, 7-9
Automobiles, 121

Backed-out trim, 54
Ballast, 75, 76
Baltic birch plywood, 66

Band saws, 153
Banks, 7, 17-21
    salvaging from, 56
Bar joists, 81
Bar levels, 150
Barnwood, 47
Barter, 43-44, 56
Baseboards, 53
Basement walls, 23
Beds, 121
    built-in, 56
    Pullman, 122
Bench grinders, 152, 153
Beveled glass windows, 106
Bidets, 136
Bits, 101
Blackboards, 68, 115
Bookcases, 56, 97
Bookshelves, 53
Boring jig, 101
Boxing boards, 45
BRD or B.R.D.,
    see Building Regulations
    Department
Brick, 115
    embossed, 30
    fireplace, 111
    used, 30
Bridge, 39
Bucket seats, 121
Budgeting, 6
Building board, 67
Building codes, 6-7, 13-15, 40
Building permit, 13
Building Regulations
    Department, 14, 17

Burlap, 112

Cabinets, 5, 57-61
    antique, 103
    assembly, 59-61
    drawers, 60, 69
    hanging doors, 59
    shelves, 60-61, 69
    shop-built vs. built-in, 58
    standard sizes, 58
Carpet, 5, 111-116
    laying, 154
    pads, 92
    stretchers, 113
Casings, 52-53
Caulk, 5, 80, 88, 105
Ceilings, 68, 109-112
    illuminated, 91
    panels, acoustical, 78
    tiles, 5
    trim, 52
Cellulose insulation, 78, 79
Celotex board, 67
Cement mixer, 151
Cement-composition shingles,
    73
Ceramic tile, 111, 115
Chairs, 119, 121
    measuring, 121
Chambers ovens, 124
Chipboard,
    see particleboard
Chisels, 101, 151

Circuit breaker box, 127
Circular saws, 152
Clay roofing tiles, 73
Clay floor tiles, 115
Co-signed loans, 20
Coat hangers, 119
Cold-storage buildings, 79
Combination squares, 151
Commercial doors, 96
Composition roofing, 72
Compressors, 125
Concrete, 23-28
    blocks, 29, 30
    floors, 115
    forms, 23, 65
    mixing, 24
    moisture, 23-24
    nails, 116
    precast, 24-26
    pre-stressed, 25-26
    reinforcing, 24
    roofing tiles, 73
    salvaging, 24
    sawing, 25
    vibrating, 24
    wood as substitute, 28, 38
Conduit, 128, 154
Conduit bender, 154
Construction finance
    companies, 19
Copper, 82
Copper wire, 84
Corian, 93
Corner blocks, 53
Cornice, 52
Coved tops, 89
Crawl space, 80, 91
Credit unions, 19
Creosote, 38
Cut stone, 31
Cutting oil, 81
Cypress, 35

■

Dead bolts, 101
Demolition sites, 9-10, 81
Dental clinics, 136
Desks, check-writing, 56
Dishwashers, 123-124
Doors, 5, 95-100
    antique, 99
    commercial, 96

dutch, 100
garage, 99-100
glass, 87
hollow core, 95
hollow metal, 98
patio, 99
pocket, 98, 103
solid core, 58, 96, 118
standard sizes, 95
stile and rail, 97-98
window cut-outs, 97
Drill bits, 152-153
Drill press, 153
Drills, 152
    electric, 81
Driveway, 23
Dryers, 126
Drywall, 109
Duct lining, 78
Duct tape, 133
Ducts, 78, 142, 146

■

Education, 21
Electric drill, 81
Electricity, 127-128
    safety, 127
    salvaging, 127
    surplus items, 128
    wiring, 127
Electronic motors, 145
Energy conservation, 18
Engineers, 81
Environmental Protection
    Agency, 14
Equipment, 21
Escutcheon plates, 101, 103
Etched glass windows, 106

■

Fabric roofs, 76
Famowood, 54
Fans, 142-143, 146
    attic, 143-144
Faucets, 136
    washerless, 136
Felt-paper, 116
Fiberboard, 68, 78
Fiberglass, 90

corrugated, 89
insulation, 78
roofs, 75
shingles, 72
Financing, 17-21
Finish hardware,
    see Hardware, finish
Fire protection, 18
Fire resistant materials, 73
Firebrick, 33
Fireplaces, 141
    brick, 111
    dampers, 103
Fireproof wood, 73
Flakeboard, 64
    see also particleboard
Flame retardation, 79
Flashing, 82
Floor coverings, 113-116
    rejects and seconds, 113
Floor slab, 23
Flooring, 46, 154
    concrete, 115
    wood, 115, 116
Flourescent light lenses, 90
Flue tiles, 33
Foam mattresses, 121
Foam roofs, 75
Foam rubber, 92
Footing trench, 23, 28
Formica, 89, 97, 135
    see also plastic laminate
Foundation, 23
Four-legged tubs, 135
Framing squares, 151
Freezers, 125
Freezing, 6
Furniture, 117-122
    built-in, 119-120
    measurements, 121
Fuse box, 127

■

Garbage disposals, 124
Georgia-Pacific, 64
Glass, 85
    cutting, 85-86
    insulating, 79, 88
    plate, 85
    safety, 86
    stained, 87
    tempered, 74, 85, 87

transporting, 85-86
  wire, 87
Glass blocks, 87
Glass bottles, 88
Glass cutting, 153
Glass doors, 87
Glue, 5
Goff, Bruce, 84
Goggles, 153
Government property, 10-12, 118
Granite, 32, 56
Grates, 103
Gravel, 28
Gravel roofs, 75
Greenhouses, 79, 92, 143
Grilles, 56, 103
Gypsum board, 109

■

Hack saw, 154
Hammers, 151
Hardboard, 69-70
  decorative uses, 70
  grading, 69
  salvaging, 69
Hardware, finish, 101-104
  antique, 102-103
  scrounging, 102
Hardwood, 35, 36
Heat exchangers, 123, 134
Heat pumps, 140
Heating, 139-142
  and borrowing money, 139
  bottled gas, 140
  electric, 140
  forced air, 140
  natural gas, 140
  oil, 140
  remote, 142
  safety, 140-141
  used systems, 139
  wood heat, 141
Heating elements, 124, 134
High-tech interior design, 118-119
Hinges, 59, 60, 101, 102
Hollow metal doors, 98
Homosote, 68, 78
Hughes, Howard, 63
Humidity, 139

■

Illuminated ceilings, 91
Industrial buildings, 10
Innertubes, 73
Insects, 79
Insulating glass, 79, 88
Insulating wall board, 67
Insulation, 27, 29, 30, 39, 68,
  77-80, 82, 84, 105, 109, 112,
  114, 123, 133, 137, 139, 153
  cellulose, 78, 79
  earth, 77
  fiberglass, 78
  materials, 78
  R-value, 77
  Rock wool, 78
  scrounging, 78-79
  styrofoam, 78
Insurance, 18, 73
Interest rates, 19
Interior design, 83
Investment clubs, 19
Ironing boards, 56

■

Jig saws, 153
Jointers, 153
Junk yards, 82, 118

■

Kettle drums, 122
Kick plates, 103
Knapp and Vogt, 60
Knee braces, 57
Knives, 152
Knobs, 102

■

Lateral fields, 132-133
Lath, 47, 48
Lavatories, 135
  wall-hung, 135
Lead paint, 46
Levels, 150
Light bulbs, 129
Light fixtures, 56, 128
Light lenses, 90,91

Lighting, 128-129
  recessed, 84
  safety, 129
  scrounging, 128
Linoleum, 115
Locksets, 101, 102
Lumber,
  hardwood, 5
  treated, 28, 40
Lumber, finish, 43-49
  exotic, 44
  locating, 43
  problem woods, 45
  scrap wood, 48
Lumber, framing, 35-42
  exposed beams, 37
  grading, 38
  locating used, 35-37
  measuring, 38, 40
  pressure treated, 40
Lumber, green, 47-48
  problems with, 47
  storing, 48

■

Magnetic catches, 104
Mantles, 55
Marble, 32, 56, 115
Marlite, 70
Masonite, 69, 97, 99
Masonry, 29-33, 115
  veneer, 29
Mastic, 113, 115, 116, 154
Mattresses, 92
  foam, 121
MDO plywood, 65, 70
Measuring tapes, 151
Medical offices, 56, 136
Metal, 81-84
  buildings, 82
  in furniture, 84
  roofing, 82
  scrounging, 82-84
Metal-clad roofing, 74
Microwave ovens, 125
Milling, 43, 46
Millwork, 43, 51-61, 55
  expense of matching, 55
  exterior, 57
  measuring, 53-54
  reclaiming, 52-57
Mirrors, 87

Moisture protection, 80
Mortar, 32, 33, 115
Mortgage companies, 19
Mortising jigs, 101
Moulding, 51, 55
    picture, 53
    quarter round, 53
Movie theaters, 56

■

Nailers, power, 116
Nails, 5, 28, 67, 73, 78, 115, 152
    concrete, 116
    finish, 46, 54
    hidden, 35, 36, 43, 45
Native stone, 31
New Shelter, 84

■

Oak, 35
Office buildings, 56
Old homes, salvaging, 45
Open-end borrowing, 19
OSHA, 47
Ovens,
    Chambers, 124
    microwave, 125
    warming, 125
Overhangs, 72

■

Packing material, 79
Paint, 5, 110
    lead, 46
Paint additives, 110
Pallets, 12
Paneling, 5, 46, 66, 96, 110
    measurement, 46
    re-using, 46
    royalcoat, 70
    staining, 46
Particleboard, 5, 66-69
    bargains, 67
    safety, 66
    sizes, 67
    strength, 67
Pegboard, 69

Percolation test, 132-133
Perlite, 78
Photoelectric cells, 145
Piano keys, 48
Pianos, 121
Picture moulding, 53
Ping pong tables, 63
Pipe, 131-133, 136, 142, 146
    plastic, 92, 132
    steel, 131
Pipe posts, 81
Pitch, 71
Pitch roofs, 75
Planers, 45, 151
Planing, 43, 47
Plans, 20, 55, 72
Plaster, 110
Plastic, 89-93
    cutting, 89
    pipe, 92, 132
    uses, 90-93
Plastic film or sheeting, 5, 112, 116
    see also Visqueen
Plastic laminate, 59, 67, 70, 89, 135
    see also Formica
Plastic-finished panel, 70
Plate glass, 85
Plexiglass, 90
    roofing, 74, 118
Pliers, 151
Plinth blocks, 53
Plumb bobs, 151
Plumbing, 131-137
    alternative uses, 131
    codes, 131
    planning, 132
    safety, 137
    salvaging, 131
Plumbing fixtures, 134-136
    acrylic, 135
    cast iron, 135
    enamel, 135
    fiberglass, 135
    installing, 136
    porcelainized, 135
    repairing, 135
    salvaging, 135
Plywood, 63-70, 109
    Baltic birch, 66
    concrete forms, 65
    exterior, 28, 64-65
    grain matching, 64
    interior, 5, 64-65
    moulded, 66

quality, 63
strength, 63
Pocket doors, 98, 103
Poke greens, 3
Porcelain receptacles, 128
Portland cement, 24
Power nailers, 116
Printing plates, 82
Pullman beds, 122
Pulls, 103
Pumps, 133, 146
Putty, colored, 54

■

Quarter-round mould, 53
Quonset huts, 82

■

Radial-arm saws, 153
Radiators, 142
Railroad ties, 35, 38, 39
Real estate, 9, 20
Rebars,
    see Reinforcing bars
Recessed lighting, 84
Refinishing, 118
Refrigerators, 125
Reinforcing bars, 24
Rigid roof panels, 78
Rock wool insulation, 79
Rodents, 79
Roll roofing, 72, 73
Roofing, 71-76, 153
    around protrusions, 76
    asphalt, 75
    composition, 72
    fabric, 76
    fiberglass, 75
    flat roofs, 74-75
    foam, 75
    gravel, 75
    measurement, 71-72
    metal-clad, 74
    pitch, 71, 75
    pitched roofs, 72-74
    plexiglass, 74, 118
    salvaging, 72
    single-ply, 75
    tar, 75

tin, 74
Room dividers, 120
Rosewood, 48
Routers, 101, 152
Royalcote paneling, 70

■

Saber saw, 152
Safety glass, 86
Sales, 7
Salvaging old homes, 45
Sanders, 152
Sash cords, 105
Savings and Loans, 18-19
Saw, 151
    band, 153
    circular 152
    hack, 154
    jig, 153
    radial-arm, 153
    saber, 152
    table, 153
    tile, 115
    wire, 32
Sawing, 43
Sawmills, 47, 48
School buildings, 56
Screeding, 151
Screeds, 116
Screen wire, 80
Screwdrivers, 151
Sealants, 136
Sealing cracks and joints, 80
Seat wrench, 136
Seating, airport terminals, 119
Seats,
    bucket, 121
    tractor, 122
Semi-truck trailers, 82
Septic tank, 132
Sewage lagoon, 133
Sewage systems, 14
Sewer, 132
Sheathing, 45, 46, 63, 68
Sheet-vinyl, 115
Sheetrock, 5, 109-110
Sheetrock mud, 109
Shelving, 65, 97, 120, 121
Shingle cutter, 73
Shingles, 72, 73, 82
    asphalt, 72
    cement-composition, 73

fiberglass, 72
    wood, 73
Shipping crates, 12, 45, 66
Shower heads, 136
Shower stalls, 90, 135
Shutters, 79, 145
Sidewalks, 23, 24
Siding, 45
Single-ply roofs, 75
Sinks, 135
    stainless steel, 136
Skylights, 90
Slate, 115
    roofing tiles, 73
Slipcovers, 118
Solar, 56, 134, 145-147
    active vs. passive, 145
    collection, 92
    collection panels, 82
    collectors, 87, 146
    heaters, 134
    hot water heaters, 147
Solar-gain coils, 134
Solid core doors, 118
Speed squares, 151
Splash blocks, 24
Squares,
    combination, 151
    framing, 151
    roofing, 71
    speed, 151
Stained glass, 87
Stained glass windows, 106
Stainless steel sinks, 136
Stairways, 54, 55
Stapler, 80, 82
Steel,
    beams, 81
    cables, 84
    drums, 83, 143
    pipe, 131
Stone, 115
    cut, 31
    cutting, 32
    native, 31
Storing materials, 5
Stoves, 124
Strike plates, 102
Stripping, 52, 118
Stucco, 33
Stud walls, 28
Styrofoam, 79, 144
    insulation, 78
Sub-flooring, 46

Sun rooms, 79

■

Table saws, 153
Table tops, 119
Tar paper, 73, 80
Tar roofs, 75
Taverns, 56
Tectum boards, 68
Tempered glass, 74, 85, 87
Termite shielding, 82
Theft, 6
Thermopane, 88
Thomas Register, 89
Tie wire, 24
Tile, 116
    ceramic, 111, 115
    clay, 33, 73, 115
    concrete, 73
    slate, 73
Tile saw, 115
Tin roofing
Tools, 21, 149-155
    air-powered, 152-153
    brands, 155
    buying used, 149
    cutting, 149
    electric, 149
    for measuring, 151
    hand, 150
    keeping sharp, 150
    metalworking, 153
    painting, 154
    plumbing, 154-55
    power, 152
    squares, 151
    wiring, 154
Tractor seats, 122
Trash compactors, 124
Treated lumber, 28, 40
Tree wells, 31
Trim, 51, 52, 55
    backed-out, 54
    window, 53
Trucks, 7
Tubes, 142, 143
Tubs, 135, 136
    four-legged, 135

■

Underground home, 77
Upholstery shops, 92
Urinals, 136
Utilities, 13
Utility poles, 40

■

Valves, 136, 146
Vapor barriers, 78-80, 91, 112
Veneer, 63, 67
Ventilation, 143
    attic fans, 143-144
Vents, 80, 144
Vermiculite, 78
Vinyl, 116
    sheet, 115
    wall covering, 111
Visqueen, 80,91
    see also Plastic film

■

Wainscot, 55
Wall-hung lavatories, 135
Wallcovering, vinyl, 111
Walls, 109-112
    basement, 23
    concrete, 30
Warming ovens, 125
Warping, 6
Washerless faucet sets, 136
Washers, 136
Washing machines, 126
Water, 133
Water heaters, 123, 133
Waterbeds, 121
Waterproofing, 30
Weatherstripping, 80
Wells, 133
Weyerhauser, 64
Windows, 5, 105-107
    aluminum, 106
    beveled glass, 106
    church, 106
    etched glass, 106
    jalousy, 107
    sash weights, 105-106
    stained glass, 106
    steel, 107

storm, 107
    terminology, 105
    trim, 53
Wire, copper, 84
Wire mesh, 24
Wire saw, 32
Wire-glass, 87
Wood, 35
    filler, 54
    fireproof, 73
    flooring, 115, 116
    shingles, 73
    substitute for concrete, 28, 38
Woodstoves, 141
    safety, 141
    salvaging, 141
Wright, Frank Lloyd, 82

YOU WILL ALSO WANT TO READ

☐ **14046 HOW TO BUY LAND CHEAP,** *by Edward Preston.* This classic book covers *all* the ways of buying land at rock-bottom prices. Where to start, what to do, buying land for back taxes, at auctions, buying tax land in Canada, inside tips, and much more. *1991, 51/2 x 81/2, 98 pp, illustrated, soft cover. $9.95*

☐ **17024 UNINHABITED AND DESERTED ISLANDS,** *by Jon Fisher.* This unique book covers more than 150 uninhabited and deserted islands in the remote regions of the world, from the Pacific Ocean to the Antarctic, from the Atlantic to the Indian Ocean. Each island is described as to history and physical conditions, and maps are included to show the exact location of each uninhabited island. *1983, 5½ x 8½, 116 pp, more than 40 maps, indexed, soft cover. $9.95.*

☐ **17037 GREAT HIDEOUTS OF THE WEST,** *by Bill Kaysing.* This exciting book tells you all you need to know to find and use a hideout in the West. This book presents ideas about the hideout concept so that you use your own creative imagination to develop a hideout that will be all that you desire. *1987, 5½ x 8½, 170 pp, profusely illustrated, soft cover. $8.95.*

☐ **17049 THE EDEN SEEKERS GUIDE,** *Edited by William L. Seavey.* If you're unhappy where you are, your options for change have been confusing. What's better? Big city, small town? Stateside, or overseas? Mountains, deserts, plains? *The Eden Seekers Guide* is a compilation of some of the leading edge thought on place choices — it can get *you* started on the road to your own personal Eden. *1989, 5½ x 8½, 169 pp, soft cover. $12.95.*

☐ **17040 SHELTERS, SHACKS, AND SHANTIES,** *by D.C. Beard.* This fascinating book ("written for boys of all ages") tells in step-by-step illustrated detail how to build any kind of shack, shelter, or shanty you might ever need or want. This is one of the most useful (and *fun!*) books ever printed. *1914, 5½ x 8½, 259 pp, more than 300 illustrations, soft cover. $11.95.*

And much more! We offer the very finest in controversial and unusual books — a complete catalog is sent FREE with every book order. Enjoy the best — from **Loompanics Unlimited!**

_____JUNK

LOOMPANICS UNLIMITED
● PO BOX 1197 ●
Port Townsend, WA 98368

Please send me the books I have checked above. I am enclosing $ (including $3.00 for shipping and handling of 1 to 2 books, $6.00 for 3 or more). *Washington residents also include 7.8% sales tax.*

NAME _____

ADDRESS _____

CITY_____

STATE_____ZIP _____

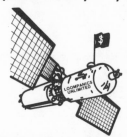